W0227386

Buddhist and Hindu
Architecture in India

Second Edition

Buddhist and Hindu Architecture in India

Second Edition

SATISH GROVER

CBSPD

CBS Publishers & Distributors Pvt Ltd

New Delhi • Bengaluru • Chennai • Kochi • Kolkata • Lucknow • Mumbai
Gujarat • Hyderabad • Jharkhand • Nagpur • Patna • Pune • Uttarakhand

**Buddhist and Hindu
Architecture in India**
Second Edition

ISBN: 978-81-239-0974-8 (HB)
ISBN: 978-81-239-0973-8 (PB)

Copyright © 1980, 2003 by Satish Grover

Second Edition   **2003**
  Reprint   2007, 2009, 2010, 2012, 2014, 2015, 2017, 2022, **2026**
First Edition   1980

All rights reserved. No part of this book may be reproduced or transmitted in any form or by any means, electronic or mechanical, including photocopying, recording, or any information storage and retrieval system without permission, in writing, from the publisher.

Published by **Satish Kumar Jain** and produced by **Varun Jain** for

CBS Publishers & Distributors Pvt Ltd
4819/XI Prahlad Street, 24 Ansari Road, Daryaganj, New Delhi 110 002, India.
Ph: 011-23266838, 23289259   Website: www.cbspd.com
                                 e-mail: delhi@cbspd.com

Corporate Office: 204 FIE, Industrial Area, Patparganj, Delhi 110 092
Ph: 011-4934 4934   Fax: 011-4934 4935
                         e-mail: publishing@cbspd.com; publicity@cbspd.com

Branches

• **Bengaluru:** Seema House 2975, 17th Cross, KR Road, Banasankari 2nd Stage, Bengaluru 560 070, Karnataka, India
  Ph: +91-80-26771678/79   Fax: +91-80-26771680   e-mail: bangalore@cbspd.com
• **Chennai:** 18/8B, Subbaraya Street, Shenoy Nagar, Chennai 600 030, Tamil Nadu, India
  Ph: +91-044-42032115, 044-26681266   e-mail: chennai@cbspd.com
• **Kochi:** 42/1325, 1326, Power House Road, Opp KSEB, Power House, Ernakulum Kochi 682 018, Kerala, India
  Ph: +91-484-4059061-65,67   Fax: +91-484-4059065   e-mail: kochi@cbspd.com
• **Kolkata:** 147, Hind Ceramics Compound, 1st Floor, Nilgunj Road, Belghoria, Kolkata-700056, West Bengal, India
  Ph: +033-25633055, 033-25633056   e-mail: kolkata@cbspd.com
• **Lucknow:** Basement, Khushnuma Complex, 7 Meerabai Marg (Behind Jawahar Bhawan), Lucknow-226001, UP, India
  Ph: +0522-4000032   e-mail: tiwari.lucknow@cbspd.com
• **Mumbai:** PWD Shed, Gala no 25/26, Ramchandra Bhatt Marg, Next to JJ Hospital Gate no. 2, Opp. Union Bank of India, Noorbaug, Mumbai-400009, Maharashtra, India
  Ph: 022-66661880/89   e-mail: mumbai@cbspd.com

Representatives

• Gujarat   0-9879558667   • Hyderabad   0-9885175004   • Jharkhand   0-9811541605
• Nagpur   0-8692091830   • Patna   0-9334159340   • Pune   0-9664372571
• Uttarakhand   0-9716462459

Printed at Sachdeva Offset Printers, Jhilmil Industrial Area, Delhi, India

For Razia, my wife

Preface

It took the reading of many prefaces to understand the meaning and intention of one. Having understood a preface to be a *raison d'être* for writing a book, it does not seem like such an onerous task. To begin chronologically, I must go back a few years to my days as a student of architecture in the Delhi School.

Though I have never been averse to travel, I really caught the bug of professional travel during a college-sponsored tour of Rajasthan, Gujarat and Saurashtra. The historic architecture of the many regions of India has since remained a constant source of nourishment for me and the fascination with India and its people an undying one. A book, then, on the history of Indian architecture seemed an ideal project for giving concrete expression to both emotions. And, finally, I hope I have managed to capture in words the first half of the subcontinent's history in the craft of building.

The format of the book is once again based on three immediate aspects of my own life—as a professional architect, as a lecturer in the history of Indian architecture, and as an urbanite with an urbanite's usual quota of social interaction. The book is aimed at being interesting reading for all the three sections of people. Courses in history of architecture in professional schools have, for better or worse, moved completely away from a detailed study of individual buildings of merely of "classified" styles to a broader understanding of architectural phenomena in their social context. The professional modern architect is also more interested in understanding the mind of the "guru-architect," the techniques of the craftsmen and the development of design ideas that resulted in the masterpieces of the past rather than in mere elaborate descriptions of individual monuments. The inquisitive urbanite too would like his monuments placed in historical perspective and need, may be, some guidance in appreciating the aesthetic achievements of the builders, rather than be overburdened with technical jargon and voluminous tomes classifying each of the millions of structures into varying styles.

The book is intended to provide sufficient reading for the student of architecture to be able to probe more detailed writings on the subject with sufficient confidence; to break down the cynicism of the practising modern Indian architect to the country's architectural heritage and to arm the more inquisitively inclined tourist of India with sufficient background material to appreciate the essence of Indian architecture. To achieve the above, it has been necessary to take some liberties from the "classic" way of writing a book on historical architecture. In this, as far as possible, I have followed chronology, for I believe that few incidents of history can be treated as isolated phenomena. In spite of the poor communication systems of the past, every human action in one part of the world is somehow derived from another, and interconnected with yet another. A study of history is meaningful only because it is so. The same principle is applicable to the history of architecture.

In the Indian context, I have emphasized this by relating the development of architectural ideas to the passage of time rather than to mere geographical locations. At the same time, one cannot gainsay the fact that due to historical circumstances there have been occasions when blossoming of a "style" has depended for its nourishment on narcissistic rather than universal knowledge. The chronological chase then has had to be momentarily given up in favour of the geographical and dynastic one. A need though to explain either phenomena against a socio-historical backdrop is common to both, for which the reader has been provided with sufficient material.

Another fact that has been emphasised and should be of interest to the reader, particularly the contemporary architect, is that the builders of the past were grappling with problems in much the same way as we architects today. They had their own brand of professional pride, their own set of "clients" to satisfy their own problems of meeting their economic needs and at the same time rationally structuring edifices to meet the "gothic" or "baroque" needs of the clients and the social times. The parallel situation I hope will be made transparent in a chapter exclusively devoted to describing the virtual "trade unions" of historical India, and the *vastupurushamandala*, that forms the geomantic, theological and academic backbone of ancient Hindu architecture.

SATISH GROVER

Contents

Introduction

Until history repeats itself and another discovery as significant and startling as that of the existence of the Indus Valley Civilization is made, any history of Indian architecture must inevitably begin from the time when the earliest known wave of immigrants settled in modern Sind and Punjab. They planned and built the now famous cities of Mohenjodaro and Harappa between 3000 and 2000 B.C. Not only are these the earliest extant remains of recognizable building activity in India, yet there is a discernible link in the traditions of the Indus Valley and that of the latter-day Hindus in spite of the most deliberate efforts of the later Aryans to decimate the urban centres of the civilization. The Brahmins of medieval India, for example, were different from the Harappan "priestkings" in so far that the temporal powers of the time had been usurped by the militant Kshatriya caste, leaving the Brahmin a mere "priest" and not king too; the bathing tank of Mohenjodaro is surely the earliest prototype of the inevitable sacred pool of any temple city of 15th century Dravida; and there is, of course, the mother goddess cult-worship that certainly carried over as Durga into the "classified list" of the pantheon of Hindu gods.

Architecturally speaking, the link is more tenuous and staccato. From the brick architecture of the Indus Valley, we are plunged into the Vedic period of bamboo and timber hamlets and only a bit more gradually into the enduring cave and stone architecture of the Buddhist and medieval Hindus. The continuity prevails, however, in the all-pervading standardisation in the techniques of building and town planning of the Harappans' and the equally inviolate Brahmin laws of theology and geometry that permeate the sculptured architectural surfaces of Hindu temples all over India. The link-up, then, with the Harappan civilization is more of the "mind" than of actual expression. The "physical connection" though is apparent from the Vedic period of about 500 B.C. right up to 16th century Hindu India.

Thus, though the forms of the thatch and bamboo huts of the Vedic village are strikingly fragile compared to the brick houses of the urban Harappan, these fragile rural forms of Aryan India were immortalised by being hewn out of enduring stone in the later periods. It is this transmutation of architectural ideas from one material to another, from one religion to another, from one people to another and from one region to another, that forms the core of this history.

No actual example of timber architecture of the Vedic period of over 3000 years ago has, of course, survived the ravages of time. Our information, though, of this period is based on the "concrete" evidence left behind by the latter day Buddhist sculptor who adorned his stone monuments with a visual relief of the legendary tales of its founder, set against the background of the Indian countryside of the Vedic period. In the renowned bas reliefs of Sanchi, for example, we see clusters of circular huts with domical thatch roofs, gable, arched timber palaces and loggias, the ornate projecting timber balconies and arcuated openings that permeated the streets of Aryan villages and trading towns with a *joie de vivre* lacking in the workmanlike and windowless streets of Mohenjodaro and Harappa.

We turn, thereafter, to the influence of the great teachers—Buddha and Mahavira—who obviously did not share in the great "joy" of the Aryan. In fact, they riddled holes into the fabric of his society by pointing out the social evils inherent in his degrading "caste system", in his tolerance of primitive Tantric rights as a means of salvation; in the glorifying of the violence of tribal Aryan wars; in its surrounding religion in an aura of Sanskrit mumbo-jumbo and elaborate ritual incomprehensible to the rural Aryan. The common Indian, then, flocked with relief to listen to the simple and straightforward sermons of these two apostles of peace, instead of being mute witnesses to the great sacrificial *yagnas* of the Brahmins.

One of the apostles, Buddha, was fortunate in finding a great patron in none other than Emperor Asoka. This fortuitous event in 326 B.C. marks the beginning of the longest and most enduring part of the history of Indian architecture—that of the use of stone as a building material. Asoka took his cue from the contemporary Persian stone masons who had already built the marvellous stone palaces of Darius and Xerxes, and set up schools of stone craftsmanship, most likely supervised by Persian gurus. The graduates from these colleges ultimately built monuments of stone to the enduring glory of Asoka's state-patronised religion of Buddhism. *Toranas, chaityas, stupas, stambhas,* and *vedikas* dotted the countryside of Buddhist India of the Asokan period. Crafted in stone as they were, these monuments are witness to the Buddhist masons' skills in the "carpentry of stone."

The Buddhist monks, however, seem to have had little confidence in the endless survival even of these stone monuments. As they retired to isolated plateaus and mountains to meditate in peace, they decided to enlarge and transform the natural grottos and caves of the hillsides into great and glorious prayer halls and monasteries for their clergy. They were now certain that as long as the gorges and cliffs of Ajanta, Ellora, and Karle survived the monuments of Buddhism too would not vanish. In carving out these "wonderous caverns of light" out of the granite hillsides, they added a new and unique dimension to the Indian architectural tradition that, for lack of more appropriate nomenclature, has been termed "cave architecture." There is no parallel to the aesthetic achievements of the Buddhist "rock architects" in any other order of architecture anywhere in the world.

India's freedom from the tentacles of Brahminism, though glorious, was shortlived. Ironically, the first victim of the Brahmins' subtle machinations were the Buddhist clergy. In deifying the "Buddha" as "saviour," whose image could be worshipped like that of Hindu gods, the protagonists of Mahayana Buddhism played themselves into the all-absorbing pantheon of Hindu gods. The Brahmin was only too keen to accept Gautama the Buddha as an *avtar* of Brahma, as long as Brahma alone was recognised as the Supereme One. The lesson of the priest kings of the Indus Valley too was not forgotten by the Brahmin. He realised the necessity of setting up an alliance with the Kshatriya's—the Aryan purveyors of temporal power—who were "raring" to go to bloody war and re-live the glorious days of the *Mahabharata,* the messages of Buddha and Mahavira notwithstanding. The rather pessimistic tone of Buddhism and the great material prosperity of 5th century India paved the path for the resurgence of 'Brahminism under the great dynasty of the Guptas. Once re-established, Hindu hegemony has not relented its hold over the majority of the Indian people for well over a millenium.

It is of course the architectural output of this great resugence that we are concerned with; an architectural tradition that, after fumbling through an early period of uncertain experimentation with the forms of a Hindu "place of worship," in towns like Aihole and Badami (in modern Karnataka), emerged confident and triumphant to build the great temple-cathedrals of Konarak, Gujarat, Khajuraho and Tanjore.

In arriving at this stage of confidence, the Hindu designer took a long and tortuous journey, the rewards of which, though, seem to justify the effort. Just like Asokan craftsmen had fashioned the Vedic timber forms into *toranas* and *stambhas,* the builder of the Hindu temple took his cue from rural folk forms as well as extant Buddhist structures. The result of the former was the *shikhara* and of the latter the *vimana.* Ultimately these two towering forms became the distinguishing features of the north and south Indian style of temple architecture. The raw masonry forms of the Shikhara and Vimana that thus emerged at Aihole, Badami and Pattadakal were ultimately refined by the skill of the sculptors of Mahabalipuram who fashioned a series of experimental dummy sanctuaries out of an outcrop of living rock. Ultimately, through interaction between the cave carvers of Ajanta and Elephanta, the Rashtrakuta and Pallava rock cutters of Ellora and Mahabalipuram, the Chalukyan stone masons of Badami and the Gupta bricklayers of Bhitargaon, emerged the crystallisation of the two mother-styles—the northern and the southern.

Though identifiable schools of art—the north Indian or Indo-Aryan and the South Indian or Dravidian (as classified by traditional writers), and the rather hybrid form of Jaina architecture emerged, the "priest architect" retained enough academic control over his Hindu craftsmen to ensure the essential design unity of the architecture of India—North, South, East or West. In order to trace and illustrate this unity, one has to bridge from one centre of architectural activity in one corner of India to another. From the *shikharas*, *vimanas* and *samosans* of Osian, Tanjore and Dilwara to the *gudha mandapa*, *nat mandir* and *pradakshina paths* of Modhera, Orissa and Khajuraho, to the *prakramas*, trefoil arches and *rathas* of Rameshwaram, Kashmir and Konarak; all built by the burgeoning feudal lords to the great glory of Shiv, Brahma and Vishnu in the plains, plateaus, mountains and hillsides of India. Until the wheel, so to say, comes full circle, once again in the region of modern Karnataka, not far from Badami, the original birthplace of the Hindu temple.

Here in the cities of Halebid and Belur, the so-called northern and southern styles are submerged once again into a single entity in the great creations of the Hoysaleswaras. The rectangular offsets of the pyramidical form of the *vimana* and the curvilinear profile of the *shikhara* are transmuted into a bell-shaped tower planted over a star-shaped *garbhagriha*. The surfaces of the productions of the master architect, Acharya Janaka, are adorned with the most profusely decorated sculptures of *devas* and *apsaras* overshadowing even the sensuous forms of the famous erotica of Khajuraho in its flourishes and ornamentation.

Hereinafter, leaving the north, which earlier in 12th century had already succumbed to Islam, we witness spasmodic artistic outbursts of the genius of the southern people, who were virtually on the run from the inexorable march of Islam. The last monumental if not innovative outcome of their talents was the legendary city of Vijayanagar or Hampi, the ruins of which confirm the glorious descriptions of it recorded by the many visitors to the city in its heyday. Thereafter, we are left only with the fortress-like temple cities of the deep south. The cities of Madurai, Srirangam, and Rameshwaram do greater justice to the perseverance and indefatigability of the Hindu builder, than to his architectural genius. Miles upon miles of corridors, halls upon halls of millions of columns, rings around rings of battlemented walls and towering *gopurams* after *gopurams* are the essentials of this "style"—if it may be called one. The progressive and assimilative genius of the Indian builder seems to have dried up under the orthodox and plebian demands of his chief client—the Hindu priest—who seems to have been preoccupied with protecting his religion and archaic social systems from the influences of Islam.

History had finally caught up with India again. Another of the seemingly endless and periodic invasions of India—this time that of the Muslims—finally and mercifully called a halt to the humdrum activity of throwing concentric walls and *mandapas* of florid columns around the ancient temples of southern India.

How the invaders tapped the Indian builders' resources in giving a new lease of life to architecture in India, by building great and glorious mosques, palaces and tombs is another story.

Buddhist and Hindu Architecture in India

CHAPTER 1

Indus Valley Civilization

More than five thousand years ago, an itinerant group of people, probably of Sumerian origin, wandered into the north-west of India by way of the MulaPass and the coastal road which runs through Las Bela and Makran and crosses the Hab near modern Karachi (*Fig 1.01*). Fortuitously, perhaps, they found their way into a breathtaking green valley, richly forested and abundantly watered by the now legendary river Sindhu. Travelling further up the river, they discovered the alluvial plains of the fabled 'land of the five rivers', Sindh, Jhelum, Chenab, Ravi and Sutlej, later known as the Punjab. To these wayfarers from across the deserts of Iran, this land was God's own gift. There was no reason to wander further; the rivers assured

Fig 1.01 Routes of entry of immigrants into India from the Iranian plateau, Afghanistan, and China.

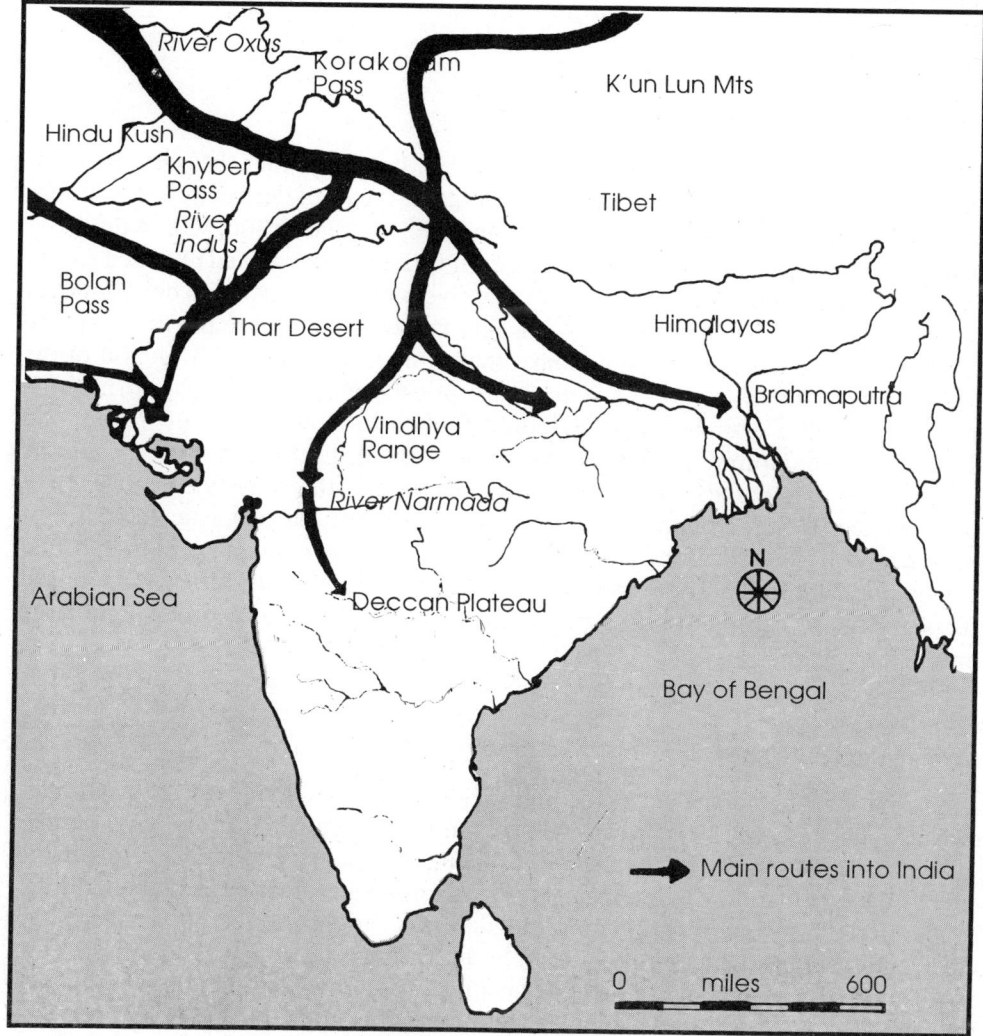

them a dependable supply of water, and the banks were ideal for cultivation. The forests around were not only an abundant source of game, but timber as well; timber that could be used as building material, as fuel for cooking and even, as they were soon to learn, for baking bricks to build permanent structures.

Over a period of almost a thousand years, the proliferating clans of these immigrants had spread themselves over an area of about half a million square miles along and around the river valley. The paraphernalia of urban life generated during this period is the substance of the earliest known civilization of India—that of the Indus Valley. Founded more than two and a half millennia before the birth of Christ, the civilization marks not only the beginning of the art of building in India, but was destined to be the crucible of Indian culture as we know it today.

During the first millennium of their settlement in the valley, the people most likely lived in clearings reclaimed from the forests on the banks of the river. Nature was bountiful to them and resources of the environment seemed inexhaustible and abundant. With their knowledge of cultivation not only of rice and wheat but of cotton as well, the community was able, probably by a system of irrigation, to produce sufficient agricultural surplus without undue strain. Such self-sufficiency in basic necessities encouraged the growth of industries like pottery, brick-making, carpentry and weaving of cotton textiles. Consequently, a prosperous mercantile class, also engaged in overseas trade, emerged. It was only natural for such a peaceful and rich community, involved in multifarious activities, to congregate in settlements of an urban nature.

Priest Kings and the City States

Inhabitants of the Indus Valley soon discovered, however, that the very elements which sustained their lives, could also become great forces of destruction. The forest, an abundant source of timber for building and fuel, abounded, too, with wild life— crocodiles, tigers, elephants and rhinoceroses. It was an unknown and untamed source of threat to human life. Otherwise life-giving rivers were also liable to devastatingly damaging floods. The common folks' meagre knowledge of these phenomena, and an equally meagre technology to combat these forces, encouraged a section of the people to set themselves up as a semi-divine priestly class. This group formed the 'magic men' who claimed that they could contain the vast forces of nature, and control the environment that appeared hostile through elaborate rituals of appeasement of the gods above. Gradually, but surely, the self-styled priests, like the latter day Hindu Brahmins, gained control over the various aspects of life of these people. They became, in fact, virtual priest kings, exercising both spiritual and temporal authority. The organization through which this authority was exercised may have been a group of peacefully coexisting city states, or even a single large empire.

Archaeological Excavations

The script used by the people of the Indus Valley, consisting of some 270 symbols, remains undeciphered. Historical processes in the region have left us with neither 'papyrus scrolls' nor a 'Rosetta stone' as a clue. Conclusive evidence is available only of the fact that they wrote from left to the right. Our knowledge of their culture is, therefore, based entirely on archaeological excavations carried out as late as the nineteenth century, which brought to light the various settlements and cities of the civilization. These, and subsequent excavations have revealed that the community had spread over a vast area. Colonies were located even as far down as Lothal in

> *Facing page*
Fig 1.02 A conjectural restoration of the city of Mohenjodaro

Gujarat State of present day India. Nevertheless, the 1,000 miles (1,600 km) long valley of the Indus and its tributaries, was the spine of the civilization. At the northern and southern ends of it stood two major seats of government. These were the similarly laid out, now famous cities of Harappa and Mohenjodaro (*Fig.1.02*). The two cities were located at the foci of an ellipse; the geographic ellipse apparently produced the agricultural surplus that sustained the dwellers of the cities. The urbanites, in turn, produced and controlled the distribution of industrial goods, and even carried on an overseas trade with the contemporary civilizations of modern Afghanistan, Iran and Mesopotamia.

The Cities of Harappa and Mohenjodaro

The planning, building and civic administration of their principal cities constitute the greatest achievements of the people of the Indus Valley. Unlike other ancient cities such as Sargom or Akkad, the towns of Harappa and Mohenjodaro were not the result of haphazard expansion of unplanned settlements. Rather, these were built as carefully planned integral units (*Fig 1.03*). Considering that this was one of man's earliest attempts at organizing his urban environment, the Indus Valley cities are outstanding examples of extremely effective town planning. For any comparable achievements in history, one would have to look almost two-and-a-half thousand years later, to the cantonment towns of the Roman empire.

Fig 1.03   A small residential block in Mohenjodaro within the grid of main and subsidiary streets

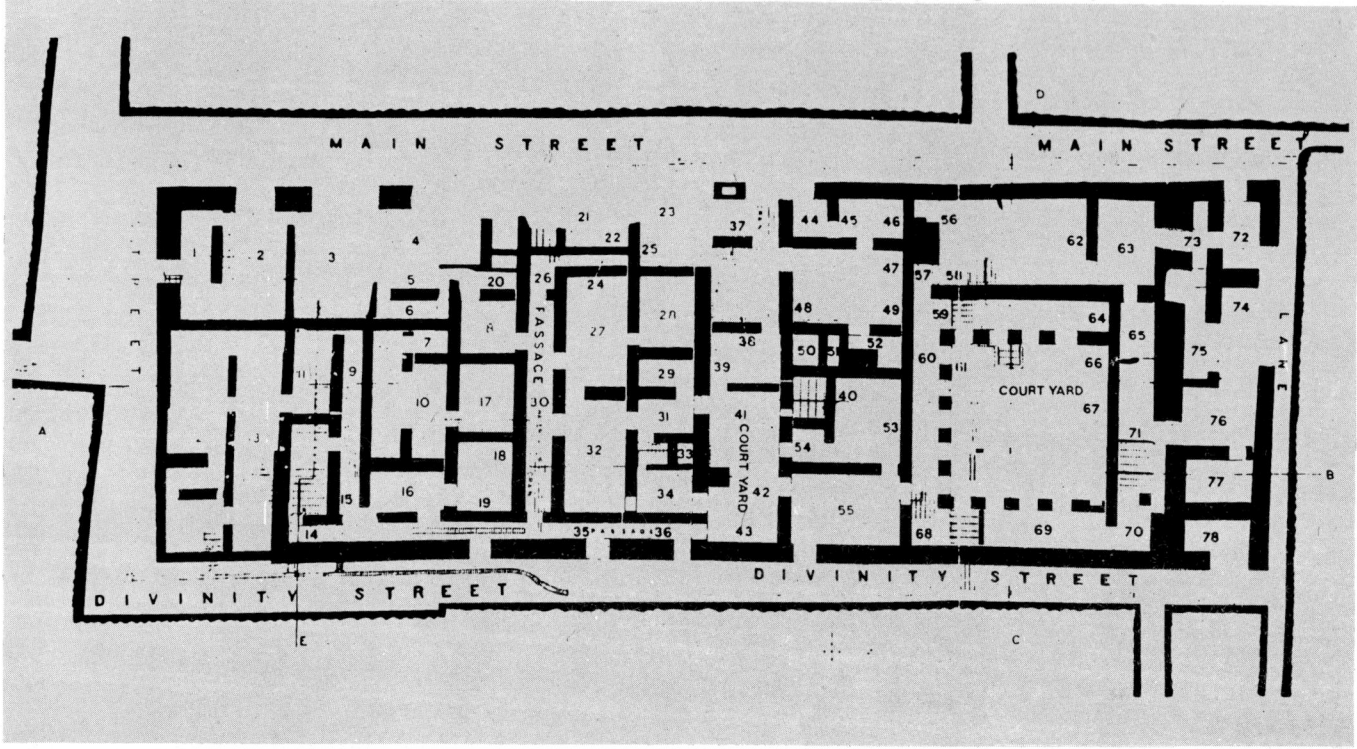

Both Mohenjodaro and Harappa were nearly a square mile in plan (*Fig 1.04*) set within cranellated defensive walls. The layout of either town was a grid-iron pattern of streets about 30 feet (9 m) wide, running in the north-south and east-west directions. The streets divided the city into 12 blocks, each measuring approximately 1200 × 800 ft (365 × 244 m). Apart from the central western blocks, the remaining ones were largely residential. The basic unit of the latter was a house of varying size and storeys, constructed with brick walls, roofed by brick tiles laid over timber rafters. It was planned as a series of rooms around an open-to-sky central court (*Fig 1.05*). From the main streets the residents approached the individual houses through irregular, narrow and shaded walkways. The residential unit invariably had no entrances opening directly to the main streets (*Fig 1.06)* and even no windows towards the subsidiary walkways. It depended for its light and ventilation entirely on the open central court, very much like most traditional housing in India even today.

Fig 1.04   Shaded area shows excavated area within the mile square layout of Mohenjodaro

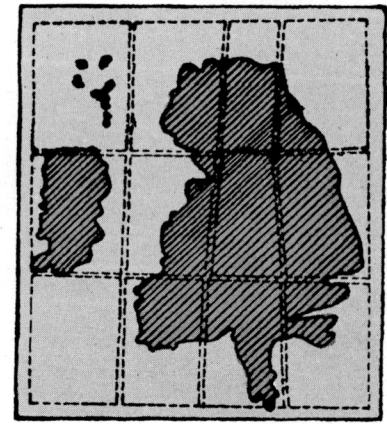

Fig 1.05 Isometric projections of a typical conjecturally restored brick house of Mohenjodaro

Fig 1.06 The typical residential unit at Mohanjodaro was built around an open-to-sky court without any windows opening on to streets

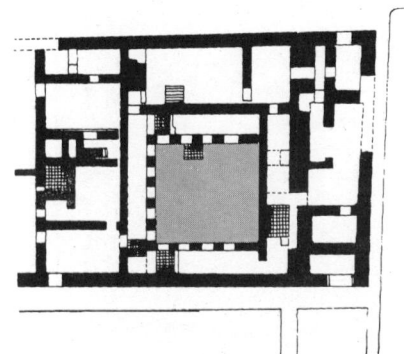

Groups of these units (*Fig 1.07*) shared a common well for their water supply, though each house had its own separate bathing area. The entire city was served by an extensive system of drainage, the bathrooms being connected to drains running under the walkways, which in turn were connected to large sewers laid out under the main street. While the smaller drains were covered with brick slabs, the larger main sewers were spanned over by corbelled brick arches. Manholes were located at regular intervals along the main sewer for inspection and cleaning of the drains. Such an elaborate drainage system took care of the abundant rainfall in the region, which at that time was not the virtual desert that it is today.

The City Citadel

For religious or other reasons, it was always the central blocks located on the western side that comprised the priestly citadel. At Mohenjodaro, it was built over a platform of baked bricks over 50 ft (15 m) higher than the general level of the city. Within it were located a palace, a bathing tank and a massive granary. Since safe storage of grain was of the utmost importance to the economy of Mohenjodaro, the granary stood on the steep verge of the citadel. Half way up its northern end, a huge unloading platform was erected. Sheaves of grain were hoisted up from here to a timber shed which was built over 27 separate massive blocks of masonry. This left sufficiently large criss-crossed ventilating ducts running at the base of the shed (*Fig 1.08*).

The citadel also had terraces at various levels approached by ramps or staircases, which were probably designed as processional paths for priestly rituals. Within the citadel are also the ruins of a vast hall measuring 230 × 78 ft (70 × 24 m) which could have been either a palace or place of worship. A tank, obviously meant for ceremonial bathing, measuring 23 × 39 × 8 ft (7 × 12.1 × 2.4 m) deep, was meticulously waterproofed with asphalt and provided with an adequate system for filling and draining. Surrounding it was a series of cells which may have been living apartments for the priests (*Fig 1.09*). The tank itself is obviously the prototype of the ritualistic bathing tanks which are an inevitable part of the later Hindu temples of medieval times (*Fig 1.10*). Close to the citadel and the granaries were located circular brick paved flour-making platforms, and a row of two-roomed barracks, apparently to house the labourers.

Fig 1.07 A typical community well in the residential area of the city of Mohenjodaro

Fig 1.08 Conjectural restoration of the granary at Mohenjodaro showing timber superstructure built over ventilating ducts and a brick masonry substructure

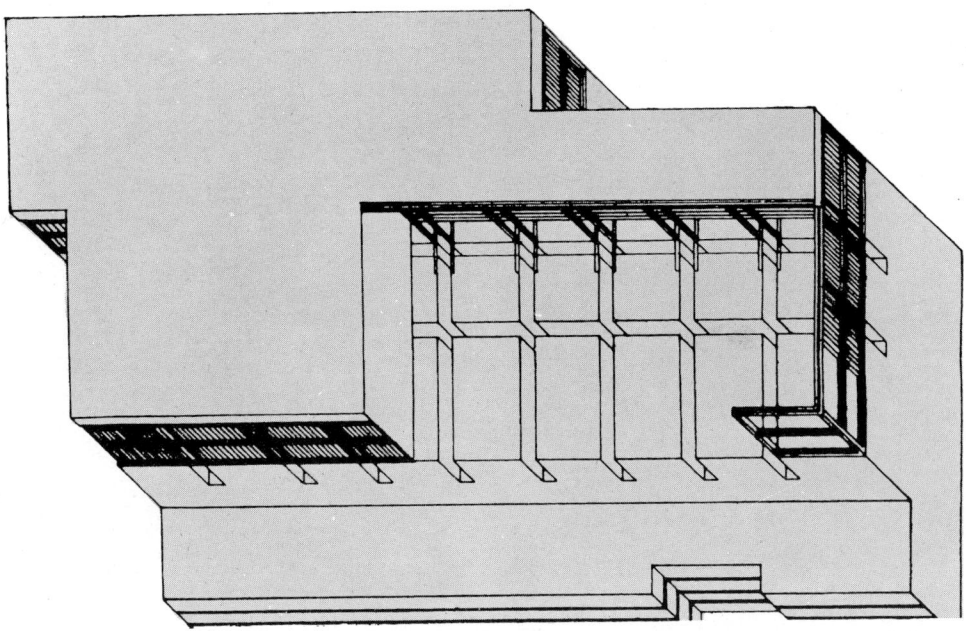

Fig 1.09 Isometric projections showing the great tank as it may have existed 5000 years ago

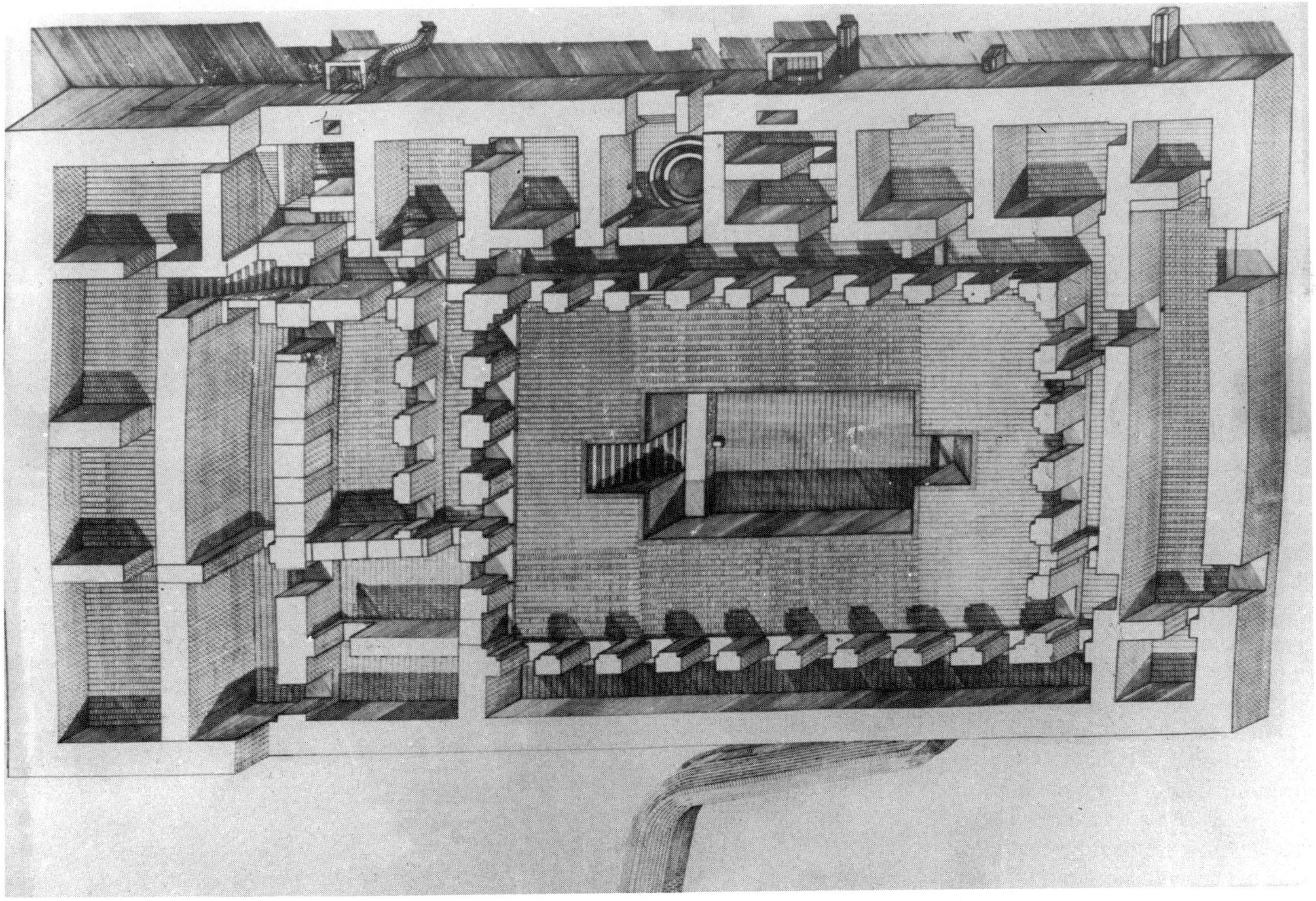

Fig 1.10 Ruins of the great bathing tank as excavated at the site of Mohenjodaro

The timber superstructure of the granaries and the monumental architecture of the citadel has virtually vanished. The houses of Harappa and Mohenjodaro have been frequently described as 'starkly utilitarian', a judgement based on the assumption that the Indus Valley people did not decorate the excavated brick shells of their houses with ornamental timber work and painting. Judging from the artistic quality of their statuary, toys, seals and jewellery, this is highly improbable. Even the empty shells of these houses indicate that the builders of the Indus Valley had intuitively evolved a system of planning for their houses, the essentials of which are eminently applicable even today to the lifestyle that is Indian.

The open-to-sky courtyard around which the house was built solved very effectively the problems of urban housing in a hot and dry climate. The central court could ventilate and light all rooms surrounding it, making openings into the street unnecessary, and thereby affording the requisite privacy. It also provided a space for open-air living within the house in which the harshness of the tropical sun was duly controlled (*Figs 1.11 a, b*). The efficacy of this planning system is proved by the fact that right up to the nineteenth century, before the advent of British rule in India, this was the accepted system of house building in Indian cities.

Fig 1.11 (a) The isometric projections of a restored house

Fig 1.11 (b) The unearthed ruins of a typical Indus Valley street with blank windowless walls of houses on either side

Secular and Religious Art

The urbane Indus Valley people were also an artistically inclined people. Though the larger part of their discovered art is of a secular nature, it is most probable that some sort of religious objects such as statues and magic symbols were made for rituals conducted by the priests. These were probably made of perishable material, and may even have been purposely destroyed to prevent these sacred objects from falling into the hands of invaders.

The remains of the secular art consist of jewellery such as earrings, pendants, belts, and bangles made in gold, silver and jade. The most interesting and also informative works of what may be called commercial art, are the numerous steatite seals. Square, rectangular and sometimes even circular (*Fig 1.12*) measuring 2–3 cm, these were cut out of steatite, and exquisitely drawn figures of animals were then sunk in with small chisels. The seals were used for security and for identification of the ownership of property. The people were also familiar with the art of sculpture. Apart from statues made of steatite (*Fig 1.13*), alabaster and limestone, images in copper and bronze were also moulded.

Fig 1.12 One of the steatite seals used for security and identification of ownership of property

Fig 1.12

Fig 1.13

Fig 1.13 A statue in steatite probably portraying one of the priest-kings

Stratification and Decay

Having developed a surprisingly high standard of building, town planning and urban living, the Indus Valley people gradually got stratified into an intensely conservative and non-progressive group. It was as though once the mould of the city had jelled, the physical fact led to stasis. Although the cities of Mohenjodaro and Harappa were destroyed several times over, they were rebuilt each time on the foundations of the earlier cities with little variation in the overall plan. All techniques and activities were governed by severe standards. The plans of cities like Mohenjodaro, Harappa, Kalibangan, Lothal and Rupar, though thousands of miles away from each other and set in different environments, are almost identical, varying only in size and detail. All the buildings of these cities were built with bricks of standard pattern and size of 5½ × 5½ × 11 in. (12 × 12 × 27 cm).

Over a period of one thousand years, there seems to have been little technological progress. Though the merchants traded with the contemporary overseas civilizations, they seem to have been reluctant to borrow any new ideas from them. In terms of personal progress they seem to have isolated themselves completely. For a long period of time they lived without experiencing any wars. So peaceful was their existence that they never felt the need to develop any kind of armaments to defend themselves against human invasions. Only the forces of nature were to be confronted and the onus for this had been taken over by the priests. The most powerful of the natural forces were the recurring floods of the Indus. These they tried to overcome partly by acts of rebuilding, and partly by erecting dams on the rivers; but largely, it would seem, by their faith in the ritualistics of the priest-kings. Gradually, their rapacious exploitation of nature and the wearisome task of repeatedly building their flood devastated cities year after year, began to sap the vigour of the civilization.

Ultimately, around 1600 BC, the cities were threatened by invaders from the west. A nomadic group of people riding horse and chariot poured into the valley and found it as attractive as it had been to the original settlers a millennium earlier. To these warlike invaders, the unarmed people of the Indus Valley were easy prey. In the face of the invasion, rituals of the priests proved ineffective and their technological isolation made meaningful military retaliation impossible. After thriving for over a millennium, Mohenjodaro and Harappa were on the defensive. During the last years of their existence, they became virtually besieged cities, under constant threat from marauding tribes. New building activity came naturally to a standstill. Additional living space was contrived by roofing over the central open courts of houses. The stately wide avenues were encroached upon not only by the extensions of houses, but even by brick kilns, which were now unsafe to site outside the city limits. The priest-kings were obviously finding it difficult to enforce the laws of governance any longer. Fully aware of the threat to their city, the rulers of Mohenjodaro were able to do little more than wall up the western gates of their city, and in the ultimate reckoning, flee for their lives. The fall of Mohenjodaro, almost three and a half thousand years ago, signalled the decay of the entire civilization.

In terms of the developmental crafts in India this was indeed a great setback. Another thousand years were to pass before the planners and builders of the invading group were able to achieve anything approaching the civil administration and town planning of the Indus Valley civilization.

CHAPTER 2

Aryan Immigration and the Vedic Period

It is fairly certain now that the hordes who subjugated the civilization of the Indus Valley were a people calling themselves Aryas, whose original home was somewhere in the legendary steppes of central Europe. Wave upon wave of Aryans in search of more favourable climes had been descending to the plains of Mesopotamia and the plateau of Iran. A group of these had for some time settled in the Iranian plateau. They were eventually driven further eastward by the pressure of subsequent immigration, and settled temporarily in Bactria, north-west of the Hindukush. Being nomadic warriors in search of green pastures, the country of Bactria did not hold them for long; soon they descended upon the primitive village settlements of Baluchistan. Hereafter, the conquest of the city states of an exhausted Indus Valley civilization was only a natural step in their march eastward.

The decimated urban settlements of the Indus Valley offered no attractions to the Aryan nomads. On the contrary, the cities seem to have aroused their contempt and wrath. Their later mythology alludes to the supreme warrior god Indra breaking the brick and earthen protective dams to destroy the cities by flooding them. The survivors of Aryan wrath on the Indus Valley inhabitants were reduced to impoverishment.

Some of them lingered on, as the *jhaluka* and *jhukar* squatters, on the remains of cities like Chanhudaro. Many seem to have migrated south to inhabit peninsular India, while a countless number may have been absorbed into the lowest social rung of the conquering Aryans. Abandoning the urban settlements of the Indus Valley to the ravages of nature, a large number of the invaders advanced further in search of richer, more virgin territories. They pushed eastward, through the natural corridor between the Thar Desert and the Himalayan foothills, and finally descended into the Gangetic plains in about 1500 BC. The richly forested plain, with its rivers and vast banks of alluvial land unencumbered by permanent urban settlements, was just what the Aryans were looking for. The natural wealth of the area had hardly been touched upon by its original inhabitants; a scattering of non-agriculturist itinerant tribes. Word of the miracles engrained in the soil of the Gangetic plain spread fast and Aryan immigration into this area began at a furious pace.

War with the Aboriginals

The Aryans attacked the "noseless and dark" aboriginals of the Gangetic plain with a ferocity no less than that of the later American white in conquering the Red Indians. With their military superiority, emphasized by horses and chariots, Aryan success was inevitable. Within a short period of time the invaders had spread themselves out in the Doab, the land between the two rivers of the Ganga and Yamuna, and foothills of the Himalaya and Vindhya mountains. The common foe, the so-called Dasyus (literally, slave) having either been vanquished or enslaved, the Aryans, like the earlier Indus Valley settler, began living in small villages, on land reclaimed from the forests on the banks of rivers. The fertility of the land of the valley of the Ganga catalysed the transformation of the pastoral habits and the economic and social organization of the Aryans; the restless nomad was gradually weaned away from his wanderings and lured into the stable and established life of an agriculturist.

The Aryan Village

Though the early Aryans had seen the use of brick in the cities of the Harappans, their descendants chose to build their village settlements in timber, bamboo and thatch which were readily and abundantly available from the forests. The chariot builder of the fighting Aryan tribes was familiar with their use and was able to adapt his skills of carpentry easily to the building of wooden structures. Timber and bamboo dwellings were also simpler and easier to maintain or rebuild in case of damage by rains and floods. The new settler was still a nomad at heart and permanent structures were against the natural grain of his existence. Moreover, to the victorious Aryans, brick structures were symbolic of a people they had conquered and whose cities and towns they held in contempt.

As depicted in later carvings on stone and described in their great literary epics, the early Aryan village was a conglomerate of timber and thatch huts of different types (*Fig 2.01*). The most elementary of the huts was circular in plan, this being the simplest to construct with bamboo and thatch. The wall was made of upright bamboos tied together with twisted twigs. The roof made with bent bamboo took a domical or conical shape, made watertight with overlapping thatch or grass. Though easy to erect, the interior of a circular hut had obvious functional limitations. An addition was then made in the form of a rectangular hut in front. In roofing this too the builders took advantage of the elastic nature of bamboo. Lengths of bamboos were bent into a semicircular shape and tied with a string at the base, much like the cord of a bow. A series of these put together over the two longer parallel bamboo walls of the hut created a barrel-like roof, which again was covered with thatch and grass. The huts were arranged in groups of threes and fours around an open courtyard. A conglomerate of such units was the typical Aryan village.

Fig 2.01 Variations of the timber, bamboo and thatch huts of the early Aryan village settlements of 1000 BC in the Ganges Valley

The village was defined and screened off from the wild life of the surrounding forest by a timber fence. To erect it, rectangular wooden posts were planted at regular intervals. Three horizontal bamboo bars were then strung between the spaces by threading them into holes in the verticals to complete the fence (*Fig 2.02*). At points of entry a portion of the fence was projected out. A gateway was installed in front of it, constructed much like the fence with its horizontal ties raised high enough to provide a controlled entry through which cattle passed to and from the pasturage. Though these village structures were obviously of a temporary nature, the forms, shaped by the early Aryan carpenter, as we shall see, contained the embryo of much of the later architecture of India, especially of the Hindu and Buddhist places of worship.

Fig 2.02 Timber and bamboo gateway installed along the village boundary wall the form of which was later transformed into the famous stone toranas of the Buddhists

Religion and Ritual of the Aryans

The worship of images was not yet a part of Vedic religious ritual and no temples appear to have been constructed. The Vedic pantheon consisted rather of Devas— 'the shining ones'—most of whom were personifications of symbols of natural phenomena like Dyaus-Pitra (Jupiter), the sky father Surya, Savitri or Mitra, the Sun god, Agni the sacred fire, and the most powerful and popular of them all, Indra, the ideal warrior. The central feature of Aryan religious life was the elaborate ritual of sacrifice to propitiate the Devas. The only architectural device required for such sacrifices was an open altar, the size and shape of which was determined by elaborate astronomical and mathematical calculations. Sacrifices on the altar could be conducted only by the learned poet-priest. He chanted hymns in classic Sanskrit in praise of the Devas, seeking their favours and goodwill for the ruling warrior aristocracy. In fact, the famous *Rig Veda* is a book of 1028 such hymns collected together probably in 900 BC. This monumental work is also the reason for this period of Indian history, between 600 and 120 BC, being popularly referred to as the Vedic period.

Administration and Economy

As the tribes settled down permanently in villages, they began to rapidly clear the forests for cultivation by burning them down. Intensive agricultural activity on the clearings led to the concept of ownership of land, and the consequent emergence of a class of wealthy landlords who had sufficient capital and leisure to trade with the surplus produce of their lands. Since the cutting of roads through the animal-infested dark jungles would have been a laborious task indeed, the rivers became the natural highways of this trade. Until now, the wars of the pastoral Aryan tribes had been fought over cattle, the accepted symbol of wealth. Now, as agricultural economy and trade became more lucrative, the Aryan tribes began to war also for control of land as well as townships which were developing to cater to the needs of trade. For reasons of security, individual tribes soon found it necessary to confederate into larger units to wage war. The civil administration of the Aryan village was left to 'samitis' and 'sabhas', analogous perhaps to a general body and an executive committee of today. The elected chief was supposed to be guided by the 'sabhas'. Gradually, however, he began to assume the privilege of a king. In course of time, by 800 BC kingship was accorded a sense of divinity by the priests, a concept that worked to the mutual benefit of both the warrior and the emerging priestly class of Brahmins, who between the two controlled the reins of religious and temporal power.

The Great Epics

The colourful religious, cultural and temporal life of this period is depicted in the epics, the *Ramayana* and the *Mahabharata*, in the form of legendary tales built probably around historic incidents and characters. The *Ramayana* describes the wanderings of the legendary Prince Rama of Ayodhya who, with his wife Sita and brother Laxmana, is banished to the forests for 14 years by a father who could not resist the machinations of Rama's stepmother. After many adventures with the ferocious aboriginal tribes of the South termed 'rakshasas', culminating in victory over the demon king Ravana of Lanka, Rama returns to his native Ayodhya. He is then crowned king and rules over an utopian state. The other epic, *Mahabharata*, is renowned for being the longest single poem in the world. It revolves around the great struggle between the hundred brothers of the Kaurava tribe ruling from

Hastinapur (near modern Ambala) and the five Pandavas with their capital at Indraprastha, most likely modern Delhi. The struggle culminates in victory for the Pandavas in the celebrated 18 day battle fought on the plains of Kurukshetra in modern Haryana. In the course of the battle, the legendary Lord Krishna delivers a message to a despondent warrior Arjuna that contains the quintessence of the Aryan and later day Hindu philosophy of life in the form of the *Bhagvada Gita*.

The Emergence of Towns

The stories interwoven into the epics are evocative of the strife for power amongst the warring Aryan clans. Ultimately, by about 450 BC four rival states had manoeuvred themselves into prominence—the three kingdoms of Kashi, Koshala and Magadha, and the republic of the Virjis. The towns of these burgeoning kingdoms and republics, such as Shravasti, Champa, Rajgriha, Ayodhya, Kaushambi and Kashi had developed into centres of industry and trade. With the establishment of large urban settlements, architecturally, India seemed to have reached a level of building and planning activity somewhat comparable to that of the Indus Valley civilization of over 1500 years ago.

Unlike the cities of Harappa and Mohenjodaro, the foundations of which were left intact once they had been deserted, most cities of the Gangetic plain have been continuously inhabited and constantly rebuilt. A number of the living cities even of today are built over such ancient sites. Thus, the archaeologist excavating in the Gangetic plain has not been able to unearth the treasures that his predecessor in the Indus Valley had been able to. It is likely, however, that the ideal town of the period was meant to be laid out as a square with a gridiron pattern. Three main streets would run north-south and another three in the east-west direction as prescribed in the *Arthashastra*, the Indian manual of town planning compiled centuries later (*Fig 2.03*).

Fig 2.03 Layout of an ideal vedic town and design of houses appropriate to the different castes as visualized by the Aryan town planners in their famous treatise, the Arthashastra

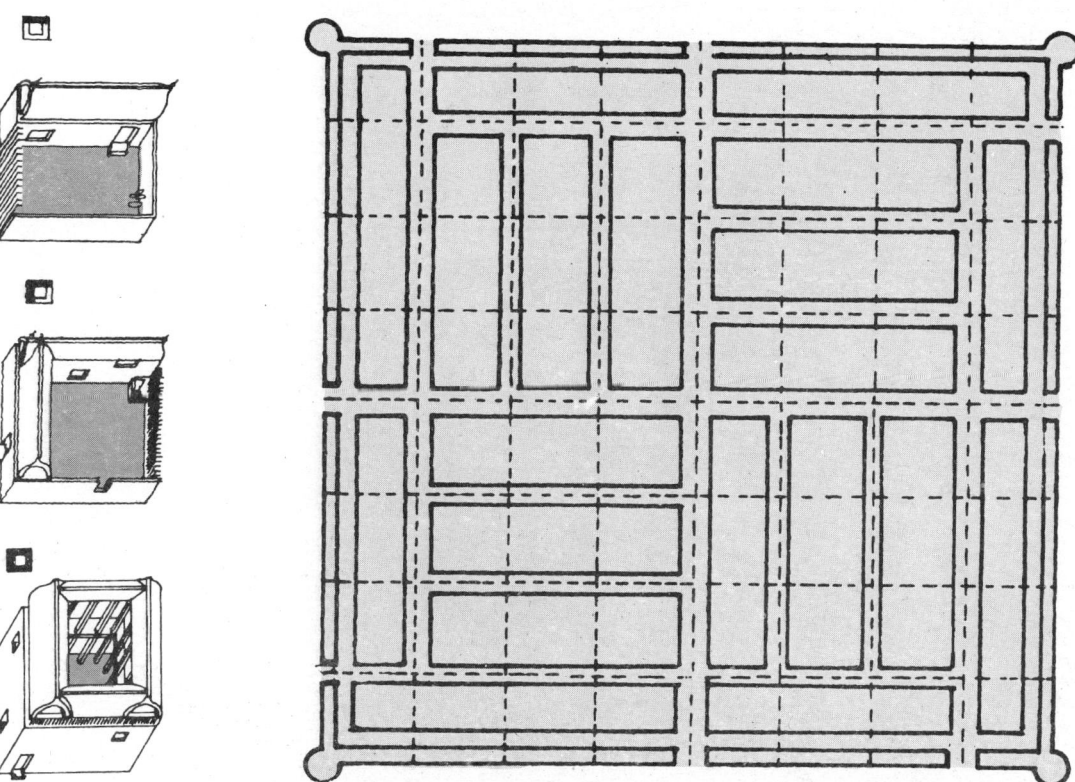

Excavations at Kaushambi, near Allahabad, have revealed that this town was built largely with baked bricks of an immense size ($19 \times 12\frac{1}{2} \times 2\frac{1}{2}$ in) ($48 \times 30 \times 5$ cm) instead of timber. Houses were built around the traditional central open-to-sky-courts. The city was protected with ramparts over 30 ft (9 m) high built in mud, riveted externally with a battered facing of big bricks (*Fig 2.04*). Rectangular towers were erected upon the bastions at varying intervals. Even the excavated site of today reveals an imposing facade. The town was also served by a rudimentary sewage system equipped with soak pits made of superimposed jars with perforated bases. Archaeological evidence suggests that Kaushambi is typical of the North Indian town during the middle of the first millennium BC. It would thus seem that by about 500 BC, after over a thousand years of settlement in India, the Aryans had established a comfortable urban way of life in the plains of the Ganga in cities such as Kushinagara (*Fig 2.05*).

Fig 2.04 The excavated remains of the ramparts of the 500 BC city of Kausambi, built with baked bricks of an immense size and rising to a height of over 9 m

Fig 2.05 *A conjectural restoration of the 500 BC gate and city of Kusingara near Magadha in Bihar*

CHAPTER 3

Buddha, Mahavira and the Mauryas

The evolution of the Aryans' nomadic ways to urbane ways of life gave birth to a new elite class of merchants, artists' guilds and bureaucrats. To this materially progressive group residing in the towns and cities Vedic religion as instituted by the Brahmins was of a rather mechanical and lifeless character. It involved the performance of rituals and repugnant bloody sacrifices that were cumbersome to the busy city dweller. The prosperous urbanities, who had worked their way up the social ladder, also resented the fact that the professional priest, apart from having abrogated all religious merit and powers to himself, was reluctant to give up the place he had appropriated at the top of the gradually stratifying caste system. This conflict between established orthodoxy and the "aspirations of newly rising groups in the urban centres" led to the growth of religious and philosophical speculations that confronted the institutionalized hegemony of the Brahmins. Some groups of non-Brahmin ascetics, having acquired psychic powers through severe penance, challenged the Brahmin claim of attributing magical religious powers only to the ritual of sacrifices conducted by him. The supporters of the atheistic philosophy of Sankhya, on the other hand, recognized the dualism of 'matter and soul' and preached that normalcy was merely the result of a balance of the three qualities of virtue, passion and dullness, the worship of the Brahmin pantheon of gods having nothing to do with it. Sects, such as the Charvakas with no mean following, dared even to preach total materialism: "When the body dies, both fools and the wise are alike cut off and perish."

Prince Sidhartha of the Shakyas

It was in such an atmosphere of speculation, inquiry and defiance of the established order that Prince Sidhartha Gautam of the semi-royal Shakya tribe renounced the pleasures and delights of a princely life that seemed to him artificial and in sharp contrast with the suffering and miseries of life outside the luxurious confines of his palace. Clad in hermit's robe he set out in search of a new path of peace and happiness. For several years Gautam studied the teachings of the *Upanishads* (commentaries on the *Vedas* by various ancient sages), the techniques of meditation, and even practised rigorous self-mortification. However, none of these showed him the way to the liberation of life from sorrow and suffering. Ultimately, after 49 days of continuous persevering meditation, in 528 BC when he was 35 years old, he found enlightenment. As Gautam the Buddha, he set the "wheel of law" rolling in his first sermon given at the Deer Park in Sarnath, near present day Varanasi. To the five disciples who heard this sermon, he preached that suffering is caused by human desire; emancipation is only possible through the eight-fold path of right views, resolve, speech, conduct, livelihood, effort, recollection and meditation. The sermon was simple and rational, not involved with complex metaphysical thinking, nor alluding to any abstract concept of God or requiring complicated rituals of worship. This became Buddha's 'middle path' to Nirvana, or freedom from the eternal wheel of birth and rebirth. To a people weary of the endless machinations of the Brahmins, Buddha's message was a welcome soothing balm. The 'middle path' over the decades was to be adopted by millions of followers of the new religion of Buddhism the world over.

Vardhamana, another scion of an oligarchic martial clan, had also, much like Gautam, set out and found his own path to salvation, 18 years earlier; his vision was to lead to the founding of another major Indian religion, that of Jainism. To Vardhamana, the Mahavira, the concept of God and the abstruse teachings of the Brahmins were of no value. "A soul resides in all things, material or otherwise." The aim of life is eternal bliss which is possible only by purification of the soul and can be achieved by living a balanced monastic life of which non-violence was the touchstone.

Kingship in the Northern Plains

The Buddhist and Jaina preaching of peace and non-violence, however, had little effect on contemporary temporal life. Rulers of the city-kingdoms and republics continued to battle ferociously for control of strategic territories. With four major contenders in the fray, the battles were now, as it were, in the semi-final stages. The kingdom of Magadha, with its capital at Rajgriha located in a valley surrounded by five hills near modern Patna, was gradually emerging as the champion power. Ultimately, India's first historically acknowledged king, Bimbisara, led the armies of Magadha to victory over the rulers of Kasi and Koshala. Bimbisara's son Ajatashatru, aided by new weapons like stone-throwing catapults and chariots fixed with knives and cutting edges, finally defeated the armies of the Republic of the Virjis in 475 BC after a 16-year war. With this decisive victory, Ajatashatru firmly established the Aryan concept of kingship in the plains of the Ganga.

The monarchial system brought its own trail of court intrigues and jealousies. Ajatashatru, who himself had gained his kingdom by killing his father, was followed by a virtual dynasty of parricides. When the last of these was overthrown, nine Nanda kings followed each other in quick succession. Their 40-year rule from Magadha was never very popular. The authority of Mahanandid, the last of the Nandas, was challenged by Chandragupta, a young chieftain to the tribe of the Mauryas. Following the astute advice of his wily Brahmin counsellor, Vishnugupta, also known to history as Chanakya and Kautilya, Chandragupta bided his time. He chose to press towards the Indus region, avoiding an immediate confrontation with Mahanandid.

Cyrus and the Trans-Indus Region

The area west of the Indus, former home of the Indus Valley civilization that had been abandoned by the invading Aryans, had remained sequestered from the development of Aryan tribal society of the Gangetic plain to a monarchial system. Cyrus, the great Persian emperor, had crossed the Hindukush mountains some 200 years earlier, and easily established sovereignty over the various disunited tribes of Kamboja, Gandhara, and other trans-Indus regions.

Takasashila or Taxila located in the present Haro Valley (near modern day Rawalpindi), the ancient capital of the Gandhara region, had inevitably become the meeting point of the Vedic and Persian learning. Though Taxila had little pretension to being a city of great architecture, its scanty remains show that there was some semblance of planning (*Fig 3.01*) in the layout of its main and subsidiary streets. The pebble-walled and plastered houses in Taxila were much inferior to the brick and timber dwellings of Harappa and Mohenjodaro. This first city, known to

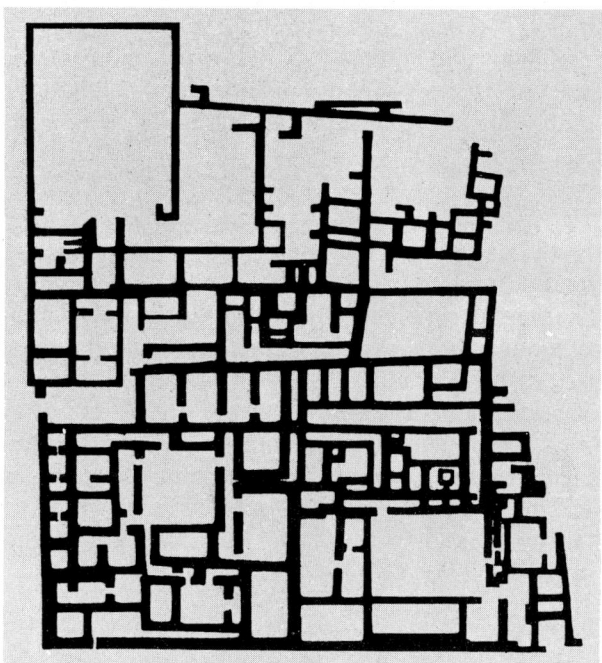

Fig 3.01 Plan of a portion of the 250 BC city of Taxila showing the haphazard growth in comparison to the well ordered cities of the Indus Valley civilisation

archaeologists as Sirkap (or the Bhirmound), was to rise to greater eminence when it was rehabilitated and rebuilt after the Greek invasion of India. In its present form it was only a living example of how the civic lessons of town planning of the Indus Valley civilization had been all but forgotten. It was at Sirkap, however, that the Persian craftsman, whose ancestors had built the great masonry palace of Darius I, and Xerxes, introduced the art of stone carving and polishing to his Indian contemporary. A most fortuitous event, indeed. As we shall see, it added an entirely new and momentous dimension to the building arts in India, which were forever destined to be dominated by the craft of stone carving and sculpture.

Alexander, Porus and Chandragupta

The Persians retained control over the trans-Indus region until their own empire crumbled under Greek pressure. After defeating Darius II in the famous battle of Marathon, Alexander's Greek armies appeared in the Indus Valley in 326 BC. His famous campaign in India lasted from February until October, during which he crossed the five rivers of the Punjab. On the banks of the Beas, however, his army, tired and exhausted after its long campaign and hard-won victory over the Indian King Porus, refused to go further. Though Alexander's ambitions were far from fulfilled, he was forced to relent and decided to return by following the course of the Indus down to the Arabian Sea. In the process he left behind a floating Greek population, and a power vacuum in the entire region. Chandragupta, the Maurya, waiting patiently in the wings, filled the vacuum admirably and was soon acknowledged leader of a confederation of northern Aryan tribes. In the meanwhile, Chandragupta's greatest ally, the wily Chanakya, had conveniently created conditions of chaos and revolt in the kingdom of Magadha. It was an opportune time for the now powerful Chandragupta to return and evict the last Nanda king. Thus, early in the third century BC and for the first time in Indian history, the plains of the two great Indian rivers, the Indus and the Ganga, came under a single great ruler—Chandragupta Maurya. Just as Ajatashatru some 150 years earlier had established the concept of kingship in India, Chandragupta Maurya could easily lay claim to being the first Emperor of India.

Chandragupta ruled his empire with an iron hand for almost a quarter century with an efficiently structured army, police and bureaucracy. His prime minister and mentor, Chanakya, minutely describes all aspects of administering and maintaining a kingdom and its many burgeoning cities, in the now famous political treatise, *Arthashastra.*

The Mauryan City of Patliputra

The Mauryan capital city of Patliputra (modern Patna) of 2000 years ago, stretched for nine miles along the Ganga, being only one and a half miles deep inland. The king's palace (*Fig 3.02*) was most likely a grand version of the familiar rectangular hut of the humble village; a series of barrel-vaulted timber roofs supported over rows of stone columns instead of brick walls. The columns were crafted either by the immigrant Persian artisans of Taxila, or Indian disciples trained in their workshops. Nothing, however, remains of these structures, except some finely jointed timber foundations excavated near modern Patna, our conjectures being based on the Buddhist craftsman's penchant for depicting scenes of Buddha's life against vivid backgrounds of the cities of the day (*Fig 3.03*).

Fig 3.02 Of the once great timber and stone Mauryan palace of Pataliputra, nothing remains but the scanty foundations also built in timber

Fig 3.03 An image of the Buddha

According to Megasthanes, the famous Greek envoy to the capital city of the Mauryan empire, the king's palace was set within beautiful parks studded with lakes, fountains, sprays, swings and artificial hillocks. The entire complex was protected by a timber reinforced wall. The city itself was a cluster of houses of varying plans invariably covered with barrel-shaped roofs, waterproofed with tiles or thatch. These, as the sculptured friezes of Buddhist stone carvers depict, had horseshoe shaped windows and lattice-screened balconies overlooking the street. These must have presented a more exuberant and lively ensemble (*Fig 3.04*) than that of the rather bleak and blank walled streets of Mohenjodaro and Harappa.

In the suburbs were the larger houses of the well-to-do, built around open courtyards, and surrounded by private gardens. Large public groves seem to have been laid out for the "recreation of man and beast." If Megasthanes's description of the towns of the Maurya period is accurate, the town planning ideas of the time seem to be similar to that of the so-called nineteenth century 'garden city' concept of Ebenezer Howard.

Fig 3.04 The Buddhist carver's depiction of the balconied and windowed streets of the Mauryan towns, that were so much more lively than the rather bleak windowless streets of Mohenjodaro

CHAPTER 4

The Ahimsa of Asoka the Great

Chandragupta's son Bimbisara, who succeeded to the throne in 297 BC, extended the limits of the Mauryan empire as far south as modern Mysore. His son, the great Asoka, came to power in 272 BC, and for some time continued the ambitious policy of expansion with as much zeal as his father and grandfather. He laid siege to the neighbouring kingdom of Kalinga (modern Orissa). After the prolonged and now famous, savage war, he was finally able to conquer the kingdom of Kalinga, eight years after his consecration as King. The hard earned victory of Kalinga, however, did not turn out to be a source of jubilation for Asoka. Rather, he seemed to have been greatly affected by the ensuing misery and suffering: one hundred and fifty thousand taken prisoner, one hundred thousand killed in war, and many more rendered homeless and dead. This, as one of his inscriptions tells us, "Asoka Priyadasi—the beloved of the gods—found very pitiful and grievous." Hereafter, King Asoka determined to atone for his sins by setting up an enlightened government, and in due course of time even entertained ambitions of becoming the "moral leader of the whole civilized world."

The religion propagated by Gautam the Buddha was tailor-made for Asoka's moral ambitions. He adopted it wholeheartedly and vigorously set about the task of propagating the message of Buddha not only within the confines of the Indian subcontinent but to the whole world. He supported the doctrine of *ahimsa* (nonviolence), substituted pilgrimages to Buddha's holy places for hunting expeditions, and even appointed so called "Officers-of Righteousness" to see that his humanitarian reforms were carried out. The first architectural manifestation of Asoka's campaign for Buddhism was to give the many nondescript earth and rubble structures that housed the relics and ashes of the Buddha, a distinguished and durable form; in fact, to transform these wayside shrines into monuments in the glory of the Buddha.

The Monumental Principles of the Ancient

Ancient monumental edifices all over the world have always been structured on the principle of being built over a wide base with the superstructure decreasing gradually in size towards an apex of sorts. To attain height, monumentality and structural stability, with essentially inelastic materials like stone and brick, this was the soundest way of assuring a long life for the edifice. A solid monument of this nature creates virtually no lateral forces unlike, say, a cubic mass that would need to be contained within well reinforced and massive walls of mortar jointed masonry. Rather, the load of the gradually tapering structure acts only downwards and as long the building material of the lowest course is not crushed by its own weight, structural immortality is virtually inherent in the design. The actual form that emerged from this sound principle acquired various shapes in different regions, influenced as it was by local geographical, geological, climatic, social and religious reasons.

It is also no wonder then that though the houses and palaces of ancient times have vanished, buildings like the pyramids of the Egyptians, Aztecs and Assyrians and the stupas of the Buddhists are still extant. The builders of Egypt for their memorials had crystallized the shape of the familiar sand dunes of the desert into great pyramids (*Fig 4.01*) or, may be, attempted to symbolize the rays of the sun from horizon to horizon, in their profiles. The Assyrian cut stepped terraces into the sides of the pyramid to create the famous ziggurats (*Fig 4.02*) more suited probably to the performance of their sacred rituals. Even in far away Mexico, the Aztecs built their temples over piles of receding masonry, designed virtually like monumental staircases to the heavens above (*Fig 4.03*). The one structural principle common to the Egyptian, the Assyrian and Aztec builder though, is that he always started from a broad base tapering to a narrow apex. Just as the ancient craftsman the world over had built his monumental edifices over a plan that was an elementary geometric profile, the evident choice open to the Buddhist architect was either a square or a circle, the two purest and most perfect of geometric forms. He chose the circle; to him not only did it symbolize the Buddhist 'wheel of law' but it was also an ideal focus for performing the Buddhist ritual of endless circumambulation of a sacred object. Moreover, its connotations were significantly contrary to the square which had formed the basis of the plans of the Vedic altars of sacrifice.

Fig 4.01 The great pyramids of the ancient Egyptians at Gizeh near modern Cairo

Fig 4.02 A ziggurat of ancient Assyria

Fig 4.03 A Mexican pyramid with "staircases to the heavens

The Buddhist Stupa

Fig 4.04 A circle in plan, a circle in elevation—the essence of the Buddhist stupa

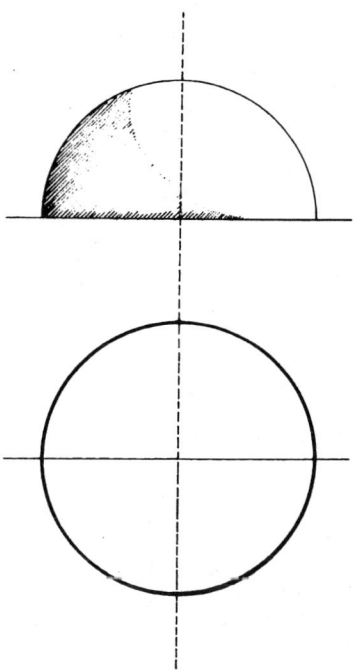

And so the irregular humble mound of rubble that had been piled over by the reverent worshippers over relics and ancient treasures to mark a sacred site, was now transformed by the royal builders into a hemispherical brick paved tumulus, the plan, elevation, section and total form of which were all derived from the circle (*Fig 4.04*). The embryo of the most powerful architectural form of Buddhism, the famous Stupa, thus emerged for the first time under the architectural patronage of Asoka.

Asoka, however, seems to have been fully aware of the transitory nature of such mud and brick stupas, however, elaborately decorated. Since his proclaimed aim was to ensure a "long endurance of the good law," he inevitably started "thinking in stone." In his meticulously planned operation, not unlike a modern advertising campaign for the spread of Buddhism, he was quick to use the well tried techniques of the Persian emperor Darius. Messages of the Buddha in the Pali text, were carved into tablets of stone and installed at various strategic locations (*Fig 4.05*). The success of his campaign can be gauged by the fact that a number of these effectively affirm the gospel of the Buddha to his followers as much as 2,000 years after they were erected.

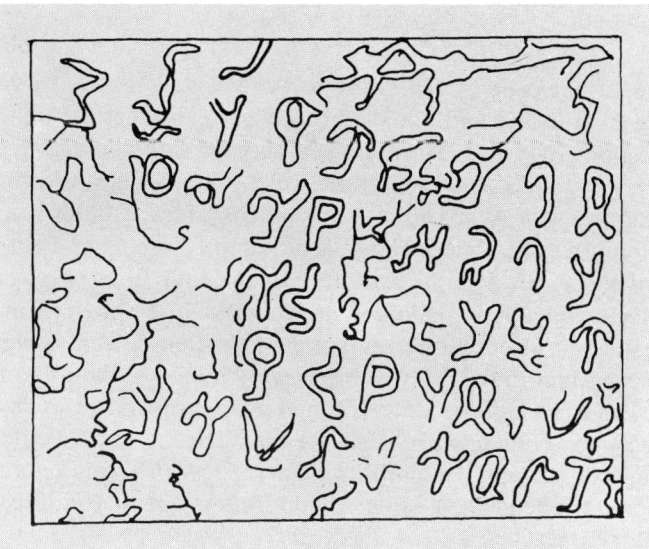

Fig 4.05 An Asokan stone-edict

The Stambhas

To these stelae, Asoka also added his own special innovation. Inspired undoubtedly by the wooden totem poles of the primitive tribes, he ordered the inscriptions carved on columns of stone instead of mere slabs, and these were set up at regular intervals along roads leading to places of Buddhist pilgrimage. The Indian stone mason had learnt his lesson well from the Persian 'gurus.' In what may be called his first test in the crafting of stone, he carried out the emperor's commands with great sculptural and technological ingenuity. Columns, some 40 ft (12 m) in length and weighing as much as 50 tonnes, were carved out from a single block of sandstone, won out of the now famous quarry at Chunnar, in modern Bihar. These massive pillars were then carried, unbroken and intact to sites over hundreds of miles away. The means of transport employed for this amazing feat are still not clear. Presumably, the operation involved building specially designed, huge bullock drawn timber carts, and even flat barges (*Fig 4.06*) for floating down the well tried highways of trade—the numerous Indian rivers.

Fig 4.06 The 'single piece' 50-ton stambhas were transported across the Indian countryside on timber carts and floated on the rivers in huge flat barges

The tapering shaft of the column, once it had reached its destination, was painstakingly varnished and polished to give it a unique and unbelievable mirror-like lustre; a fantastic achievement indeed with the sedimentary rock-like sandstone. The column, it would appear, sometimes rose straight out of the ground without any base, like the trunk of a tree, or at some places had a circular brick platform at its base. At the top was mounted a large sculptured figure of an animal.

One of these, the famous sculpture of the four lions (*Fig 4.07*) has acquired great popularity since being adopted as the national emblem of modern India. The sculpture executed in a style reminiscent of Persian design, weighing five tonnes and about seven feet in height, was effectively joined to the pillar by a two feet (60 cm) long cylindrical copper dowel, toned accurately into the shaft and sculpture without use of any cementing material. A magnificent, large as life lion sitting high in the air (as at Lauriya Nandgarh in Bihar) (*Fig 4.08*) proclaiming the *dharma* to the world, must indeed have been a wondrous sight to Buddhist pilgrims of the second century BC.

➤ *Facing page*
Fig 4.07 The famous Buddhist sculpture of the four lions from Sarnath

*Fig 4.08 The Laurya Nandgarh
lion sculpture proclaiming the
"dharma" to the Buddhist
world of Asoka*

The Ajivikas and Cave Architecture

Though Asoka zealously propagated Buddhism virtually as the state religion, he was not entirely intolerant of other sects. The Ajivika ascetics were another of his beneficiaries. He ordered sanctuaries to suit their needs to be built at the cost of the state. As if to prove the basic tenets of their thinking, that "all change and movement were illusionary" and the "world in reality was eternally and immovably at rest," the Ajivikas discarded the conventional timber and brick structures of the materially engrossed oppidan. They chose instead to carve habitable caves out of the "eternally immovable" hillsides for their sanctuaries. Also, in choosing this novel form of architecture, they intuitively immortalized the traditional dwelling of the Indian sage, who, as a mark of the rebellion from the materialistic creed of the city dweller, had for years sought refuge to meditate in natural caves in hermitical hillsides. Unknowingly, the Ajivikas had pioneered an art form that was to flower into some of the most unique works of Indian architecture.

In the shaping of these caves, the builders with chisels and hammers as their main tools, were practising more the craft of sculpture than of architecture. They were unhindered by the specifics of calculating spans, adequate supports and checking the strength of materials. If they had really chosen to exercise their freedom and thrown to the wind all the normal restraints of traditional building forms, they could well have produced a hundred bizarre and exotic shapes. However, an inherent love, or weakness for the familiar, made them reproduce in rock almost exact copies of existing structures in wood and thatch.

This curious and novel adaptation of the carpenter's art by the stone carvers, however, produced its own inimitable results that can be seen in some seven caves of the Ajivikas in the hills of Barabar and nearby Nagrajuni, north of modern Bhubaneswar in Orissa. Of these, the cave popularly known as Lomas Rishi is surely the crowning glory. Its interior circular cell 19 ft (6 m) in diameter has a hemispherical domed roof and is approached through a 33 ft (10 m) long tunnel-like hall (*Fig 4.09*). The barrel-vaulted rectangular thatch hut and the beehive roofed shrine crafted by the Aryan carpenter, found in innumerable Indian villages, had thus been perpetuated in the form of a cave.

Fig 4.09 Caves fashioned in plan, elevation and section to resemble thatch and timber huts of the Aryan village

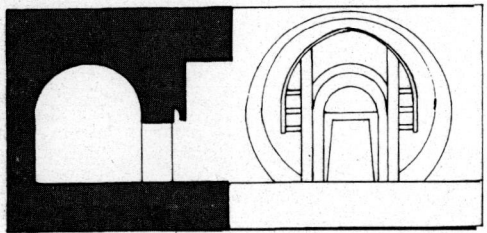

The shape of the entrance arch on the hillside is the familiar gable end of the same rural dwelling described earlier (*Fig 4.10*). The minutest structural details of the original hut have been faithfully reproduced in the Lomas Rishi. The rockcutters, however, lifted this work beyond mere plagiarism by deftly introducing a beautifully sculptured semicircular panel at the crown of the arch depicting two rows of elephants walking towards a stupa. Such a panel could neither have been lifted nor inspired by any timber structure; only a skilled stone carver could have conceived of it and created it (*Fig 4.11*). This combination of traditional building forms and applied sculpture obviously appealed immensely to the flowering genius of the Indian artist. It became the unique essence of all Buddhist and Hindu architecture in India for centuries to come.

For a considerable time after Asoka's death in 232 BC, well up to the end of the Mauryan dynasty, Buddhism for all purposes enjoyed the status of a state religion. The third Great Council of Buddhism was held in Pataliputra under the patronage of Asoka to iron out the widespread differences between the various ever-proliferating orders of Buddhist monks. Buddha may have propounded only a doctrine to help his followers on the path to salvation. His teachings, however, had generated a large international following—a following so large that it evolved into one of the earliest established religions of the world. In the country of its birth, however, it was at loggerheads with the contemporary thinking of the Brahmins, Jains and Ajivikas.

Facing page
Fig 4.11 One of the earliest sculptural embellishments applied by stone carvers to caves inspired by the traditional timber forms—the entrance to the Lomas Rishi cave

Fig 4.10 The Lomas Rishi carved out by the Ajivikas in 250 BC was the first step towards the creation of the great art of cave architecture of the Buddhists

The Cult of the Chaityas

In the process of maturing into a religion, Buddhism borrowed little from the sacrificial rites of the Brahmins; rather, its simple ritual was inspired by the popular cult of the *chaityas* (from the Sanskrit *chita*—a pyre) or 'sacred spots'. The traditional *chaitya* was a grove of trees in the middle of which small tumuli of earth had been built up over the ashes of the tribal chiefs. The rural folk found these *chaityas*, located on the outskirts of their villages, less expensive and more convenient to worship than the great Aryan gods. The Buddhist monks merely transferred the attention of their followers from these to stupas that carried in them ashes or relics of the Buddha.

With the growth of a large lay following and the assured patronage of the elite and rich mercantile classes, communities of Buddhist monks, instead of wandering around as alm begging mendicants, began settling down around such *chaityas*. The *chaityas* thus grew into small monasteries and the cult of the *chaitya* became the touchstone of Buddhist ritual and worship. By the time of Asoka's death, the Indian countryside was probably dotted by a number of such Buddhist settlements.

These establishments, to begin with, were of a very modest nature, much like the wayside village shrine of today. The stupa at best was a white-washed mound, probably decorated with festoons and prayer flags. A protective wooden fence around the stupa, apart from symbolizing the sacred nature of the enclosed area, demarcated a circumambulatory path along which monks would walk, chanting Buddhist *slokas* or verses from the holy scriptures. Near the stupa were probably some simple thatch huts for the resident monks. But for the very temporary nature of their construction, none of these settlements have survived to date. Curiously enough, the more permanent and enduring architectural masterpieces of the Buddhists were destined to be constructed after the patronising umbrella of the Mauryan dynasty vanished after the death of Asoka.

CHAPTER 5

The Shungas, the Merchants and the Buddhists

The long reign of Asoka was undoubtedly a benevolent one. Nevertheless, it was an autocracy, even if a religious one, or more precisely a Buddhist one. Brahmin discontent had been constantly simmering beneath the tight administrative organization of a centralized Mauryan authority. Gradually, the pivot of this authority had become the figure of Emperor Asoka himself. Nor surprising, then, that the power of the great Mauryan dynasty was swiftly eroded by repeated rebellions of the malcontent Brahmins, within fifty years of the death of Asoka. The spearhead of this rebellion was Pushyamitra, a Brahmin of Kanauj in central India, and a general in the army, who murdered the last Mauryan king and heralded the dynasty of the Shungas in 185 BC. The Shunga kings, for some unknown curious reason, did not take any regal title but were content to style themselves as 'Senapati' or General. This dynasty of Generals with its war-like postures paid little heed to the peaceful messages of the Buddhist religion. Inevitably, with the revival of the rule of Brahmins and their Kshatriya commanders, the Aryan rites of Vedic sacrifice were once again popularly patronized. The elaborate Ashvamedha Yagna, whereby an Aryan chief not only proclaimed suzerainty over a tract of territory but threw a gauntlet at any other contender, was performed by the Shungas, signalling the return of true orthodox Brahminism to the land if its birth.

The Growth of Sanchi

In such an intimidating atmosphere, the order of Buddhist monks began to retire from the centres of urban power. They, however, continued to enjoy the patronage of the wealthy and sophisticated mercantile class that had little to gain by servility to the rising powers of resurgent Brahminism, since their place in society was virtually preordained. Funds continued to pour in to help the order of Buddhist monks to build their new monasteries in more peaceful and solitary environments. The largest and most famous of these blossomed upon a hilltop in Sanchi (near modern Bhopal) close to Vidisha, the new centre of Shunga power (*Fig 5.01*). Sanchi had already been marked as a sacred spot when a semi-circular brick mound some 70 ft (21.3 m) in diameter had been built here, as a part of Asoka's historic architectural campaign. The stupa was now enlarged to double its size (*Fig 5.02*) by building another stone-faced mound, increasing the diameter of the original to 120 ft (36.5 m) and its height to 54 ft (16.4 m). Sixteen feet high from the ground level of the stupa, the builders created an elevated *pradakshinapath*, perhaps reserved for the clergy, the traditional one at ground level being open to the common pilgrim. The crown of the hemispherical mound was flattened out to make place for a circular platform, from the middle of which rose a three-tiered stone umbrella, set inside a square enclosure, marked out by a low stone fence.

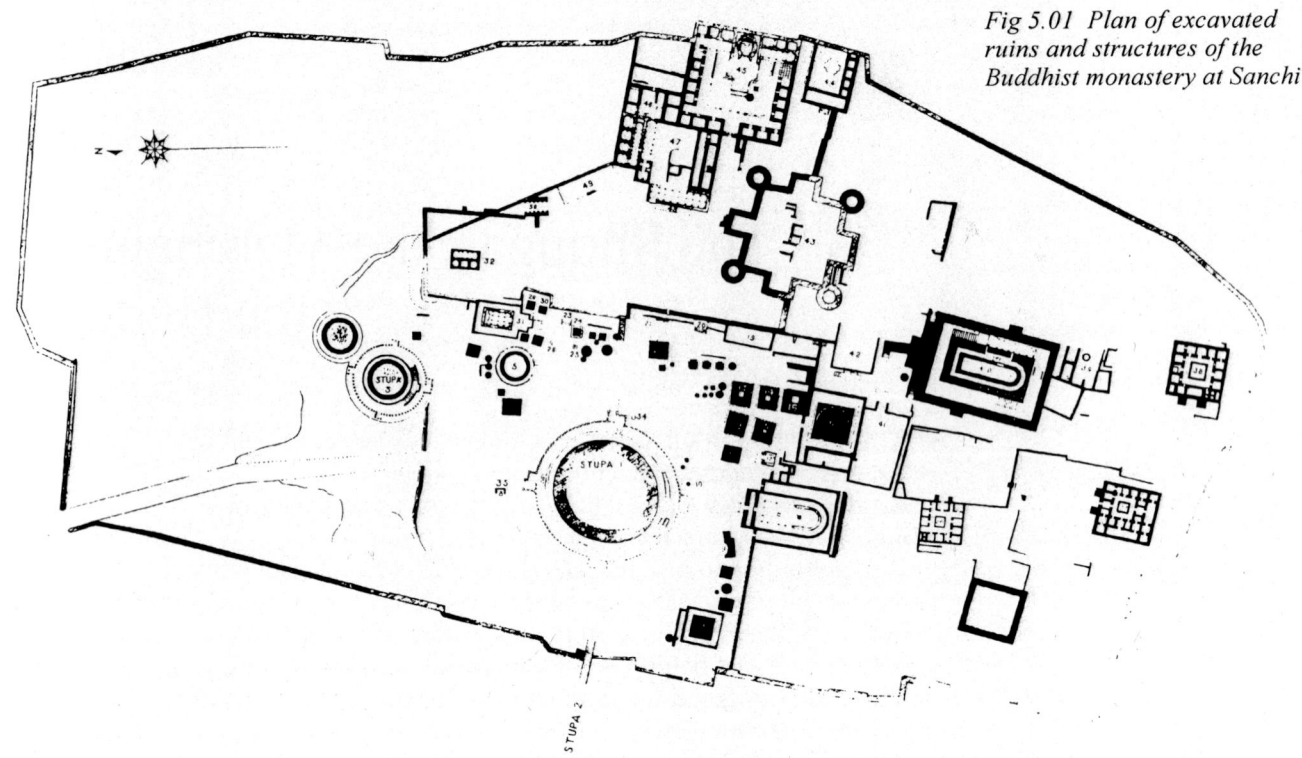

Fig 5.01 Plan of excavated ruins and structures of the Buddhist monastery at Sanchi

The Stone Vedika

In the process of carrying out the enlargement of the stupa, the traditional timber fence surrounding the original stupa had to be pulled down. It was now replaced by a massive and austere railing, fabricated entirely out of stone (*Fig 5.03a*). In assembling together this railing the craftsmen of India were taking yet another decisive step towards the proficient use of stone as a comprehensive building medium. They were already pastmasters in the sculpting and polishing of stone and had successfully installed massive Asokan pillars all over India. Now the masons, though diffident, were ready to carry out their first experiment in the use of stone as a semistructive material. No wonder then that they went about their task in a truly ponderous fashion. Octagonal stone pillars 18 in (45 cm) in diameter and 9 ft (2.7 m) high were planted along the circular periphery at a distance of about 2 ft (60 cm) from each other. The space between each pair was spanned by three 2 feet (60 cm) deep, lens-shaped horizontals, which were meticulously tenoned into corresponding mortices cut in the pillars (*Fig 5.03b*). The railing was topped by an appropriately massive stone coping, giving it a total height of 11 ft (3.3 m). It is obvious that the shape and form of the security-oriented timber fence around the Aryan village, a

◁ *Facing page*
Fig 5.02 The largest "heavenly dome" at Sanchi which has a diameter of 36.5 m and rises to a height of more than 16 m

▷
Figs 5.03 (a) & (b) Ponderous octagonal stone pillars were planted close together and spanned with lens-shaped horizontals to replace the original impermanent timber vedika around the stupa

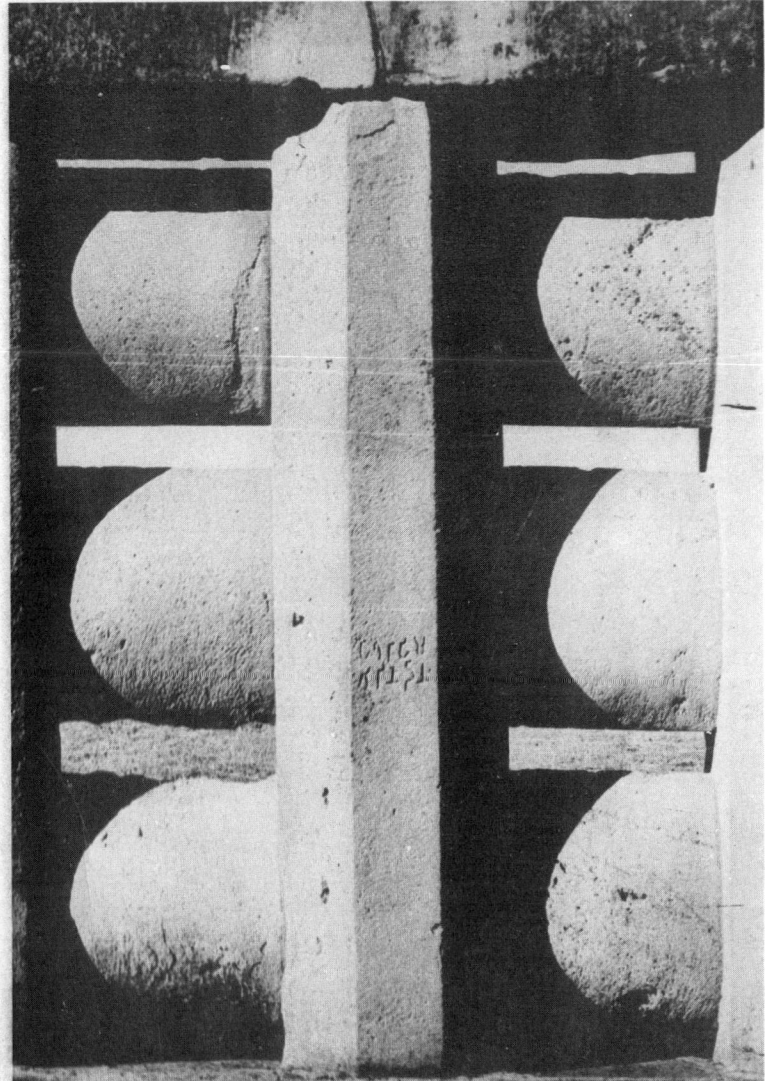

prototype of which had surrounded the early Buddhist stupa, had over the centuries also acquired sacred and magical virtues. Thus, in the entire process of erecting the sacrosanct *vedika* in stone, the craftsman had diligently reapplied the structural techniques of the Aryan carpenter so as not to tamper with the physiognomy of the timber original. Constructed to a different scale and infused with the tenacious vitality of the new material of hewn rock, this railing alone created an awesome wonder (*Fig 5.04*). The raw visual power of its architectonics is no less than that of the primitive megaliths of Stonehenge in England and Zimbabwe.

Fig 5.04 Transformed into rock, the original fragile timber vedika suddenly acquired the awesome and wondrous qualities of megaliths of Stonehenge and Zimbabwe

The Buddhist Torana

The work on the extensions and refurbishing of the stupa went on for almost 100 years. The imagination, religious zeal and vigour of the builder, as well as the coffers of the mercantile patronage, however, seemed inexhaustible. The designers were soon devising schemes to add new dimensions to the architectural ensembles. The wondrous pathway around the stupa was like a forbidding wall from outside. A sense of invitation was lacking. What could be more appropriate than to install gateways (*Fig 5.05*) to match the magnificence of the *vedika*. Elaborating once again on traditional village forms, the rural bamboo gateway of the Aryan village was transformed by the Buddhist builders into a great 34 ft (10.3 m) high *torana* (from Sanskrit *tor*—a pass). Four such gateways were installed at cardinal points of the railing. The techniques of stone masonry employed in constructing the *toranas* were no doubt as primitive as those in erecting the railing. Nevertheless, after erecting the two stone uprights the stone mason realized the futility of cutting mortices and making tenons in stone. Instead, for the first time, he applied the rationale of building with stone and merely spanned the twenty feet between the verticals with curved stone beams resting firm and square and sailing over on either side of the uprights (*Fig 5.06*). The horizontal spaces between the beams were then filled in with vertical uprights to create a sort of stone trellis, the *jharokas* of which were adorned with sculpture. The surfaces of the two pillars as well as the cross beams along the top received a generous measure of the carver's chisel and were covered with finely wrought sculpture, rich in the symbolism and imagery of Buddhist lore. It is in fact the sensuous sculpture on these gateways that has provided the Indian historian with an invaluable visual record of the myths, legends and social mores of the day.

Fig 5.05 The stupa, surrounded by vedika with gateways at the four cardinal points

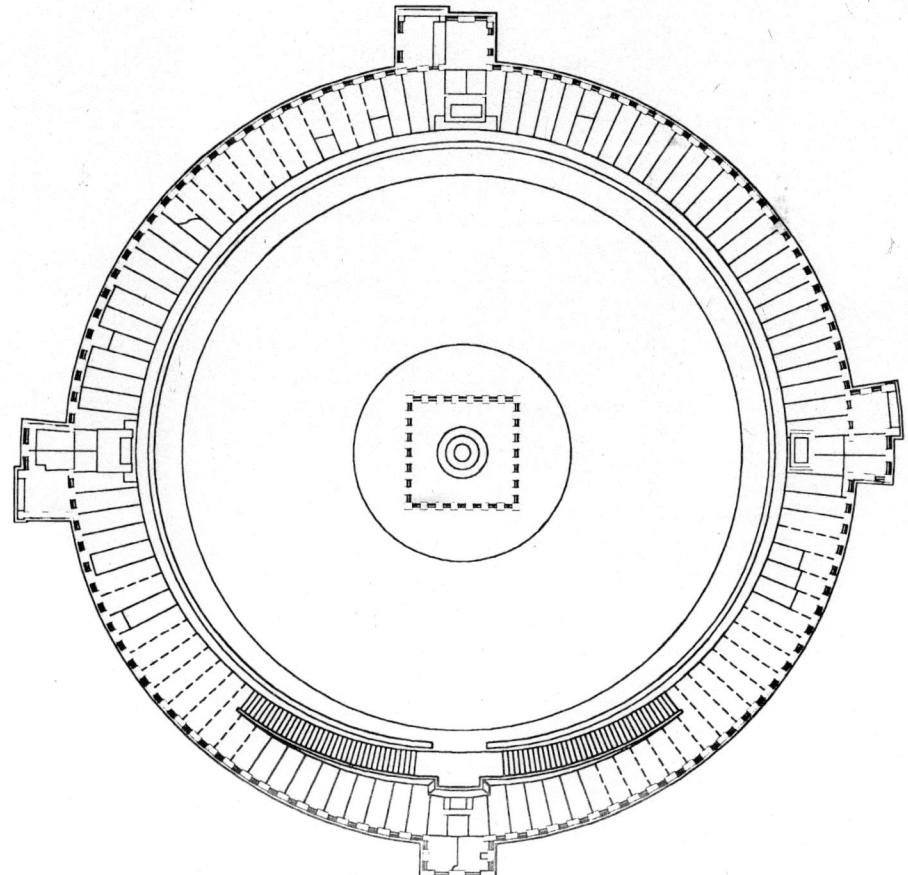

Symbolism and Structural Function

The final tectonic composition comprising the dome of the stupa (or the *anda*) the square railing at the crown (or the *harmika*) and the three tier umbrella (or *chhatra*) have in course of time been invested with various elaborate geomantic, theological and philosophical meanings. The hemisphere is the 'dome of heaven' or 'the fixed cosmic structure', the *harmika* 'heaven of the 33 gods,' the umbrella 'the world axis'. The plan of the *pradakshinapath* at the ground level is read to be a *swastika* and pregnant with the symbolism of ancient solar cults. The structural reasons for the development of such forms are, however, much more pertinent. The hemispherical stupa, as explained earlier, was a geometric crystallization of the rudimentary mound of earth eminently suitable for the ritual of endless circumambulation (*Fig 5.07*).

Fig 5.07 What started as a pile of rubble over sacred remains, was gradually crystallized into the spherical stupa

The *harmika* and *chhatra*, are a stylized visual depictions of the famous Bodhi tree surrounded by the sacred *vedika*—or railing. The *swastika* is most likely the fortuitous result of solving a planning problem. The *torana*, erected after the railing was completed, was planted some feet away from the railing. The location of its entrance was staggered from the opening in the railings so as to ensure privacy for the pilgrim circumambulating the *pradkshinapath* (*Fig 5.08*), a planning principle effectively employed later in the entrance gateways to fortresses for ensuring security. Seen purely in plan, the metaphysical observer would of course read the *swastika* in its ensuing outlines.

Fig 5.08 A staggered gateway to ensure privacy for the pradakshina path or to intentionally chalk out the sacred Swastika in plan

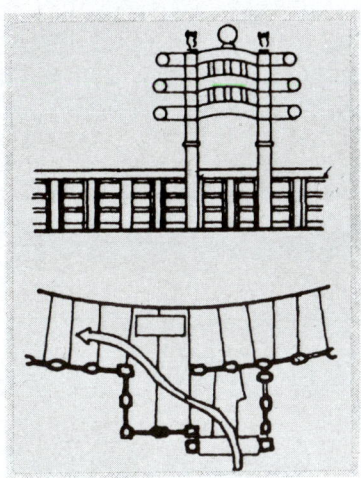

The Buddhist Vihara

An ensemble as glorious as the great stupa inevitably led to the growth of other ancillaries. Buildings were needed to house the resident monks. In contrast to the richly sculptured gateways of the stupas, the places of residence in their bareness reflect rather the inherent austerity of monastic life. These were built as a series of individual cells or dormitories enclosing a rectangular or square open-to-sky court. The open court served all the community facilities, at places including a well for drinking water (*Fig 5.09*). The cells, on the other hand, afforded the monks sufficient privacy for the practice of meditation. As one can see the basic idea was only an enlargement of the traditional concept of the 'house around a courtyard' harking back to the days of the Indus Valley civilization. In due course, these came to be known as *viharas*. Since they were built in one or more storeys, in timber frame and brick walls, only the foundations of these at Sanchi have survived, the superstructure having vanished soon after the monasteries were deserted.

Fig 5.09 Remains of one of the many courtyard viharas that provided residence for the Buddhist monks of Sanchi

Chaitya Halls

The great stupa, though a magnificent structure, had its limitations. It was essentially an open air edifice incapable of being used in inclement weather. The need was felt, therefore, for an enclosed hall in which a miniature stupa (so far the only acceptable symbol of Buddhist veneration) could be conveniently worshipped the year round. The simplest solution was to place the stupa at the end of a long rectangular hall (*Fig 5.10a*). Soon, however, the designers realized that the circular stupa sat rather incongruously within its rectangular enclosure. The walls directly behind the stupa were then made semicircular to echo the profile of the stupa (*Fig 5.10b*).

The path for the ritual of circumambulation around the stupa was also thus clearly defined. The roof of such a structure was the familiar barrel vault in timber, covered with tile and supported on brick walls framed by timber pillars. The entire composition was built on a high plinth enclosed by the inevitable sacred railing. The nascent form of the famous apsidal ended halls of worship, essentially like Roman basilicas in plan, thus emerged in the so-called Temple No 40 of Sanchi (*Fig 5.10c*), a form that was to captivate the obsessions of the Buddhist builder, though in another medium, and result in the famous *chaitya* caves at Ajanta, Ellora

Figs 5.10 (a) A makeshift rectangular hall for the stupa, (b) duly transformed into an apsidal hall to echo the circular stupa, and (c) ultimately roofed over with a barrel-vaulted and tiled roof to create the earliest permanent chaitya halls

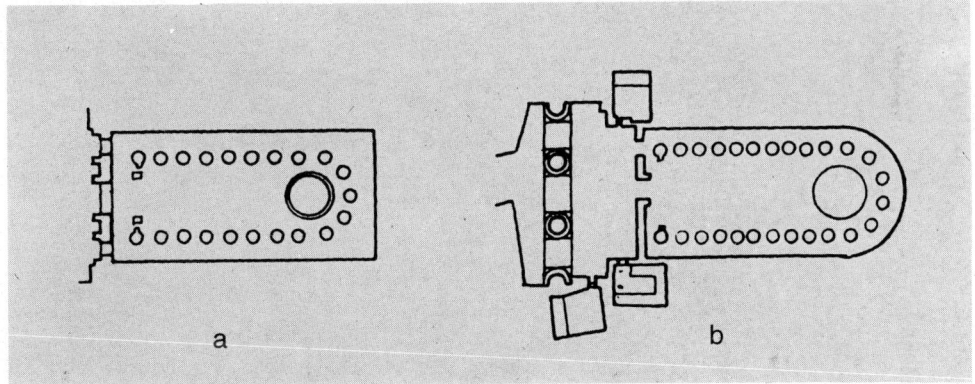

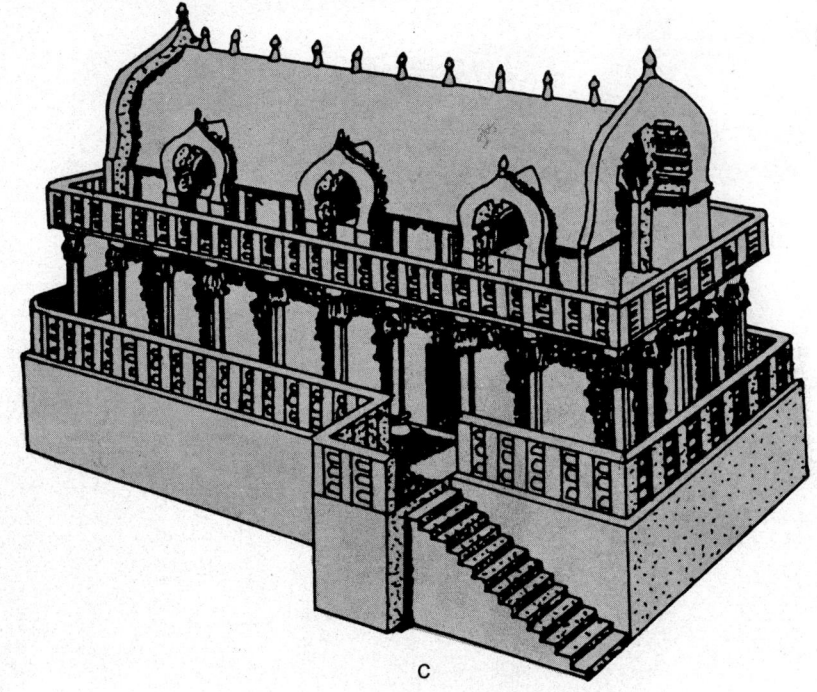

and Karle.

A congregation of such *chaitya* halls, *viharas*, votive stupas and the inevitable Asoka pillars, dispersed on terraces at various levels around the great white stupa on the hill of Sanchi (*Fig 5.11*) must have been an awe-inspiring sight to Buddhist pilgrims, about the time of the birth of Christ. For the Buddhist architect and craftsman, it represented a century of building endeavour, during which he had evolved the essential principles of design of forms that were destined to be the eternal architectural symbols of the religion of the Buddha: the *vihara*, the *chaitya*, the *vedika*, the *stambha* and the stupa. These, as we shall see, were elaborated upon for hundreds of years by craftsmen not only in distant parts of India, but in the various nations that adopted Buddha's middle path as their way to salvation.

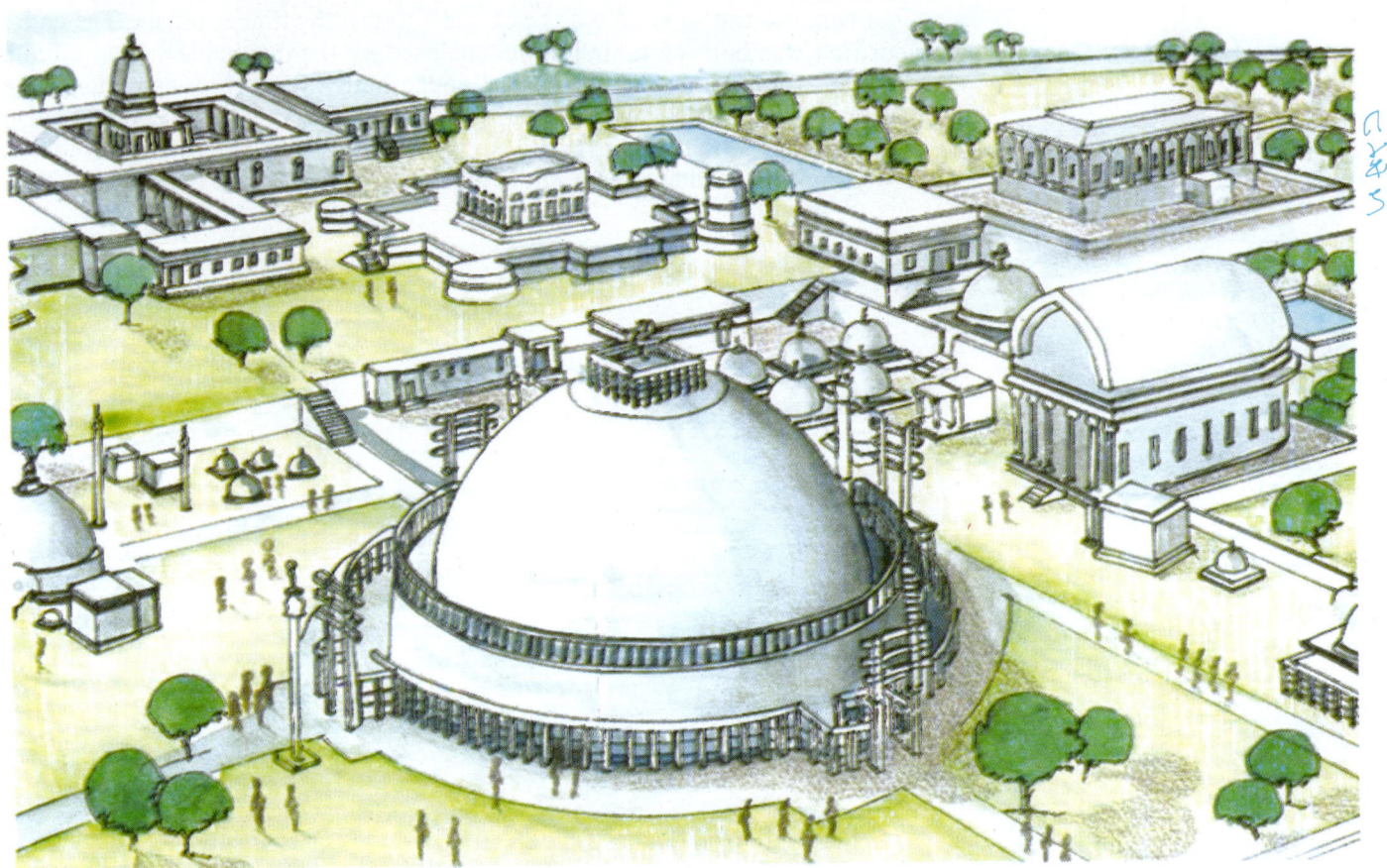

Fig 5.11 *The monastery of Sanchi as it must have been in its heyday almost 2000 years ago*

CHAPTER 6

The Andhras of the South

While Buddhist monasteries like Sanchi flourished under the ever benevolent mercantile umbrella, the Shunga kings weree hard-pressed to preserve the empire bequeathed to them by the Mauryans. As early as 206 BC the Selucid Greeks had broken down the defences of the north-western borders of the Mauryan empire by defeating the Indian king, Subhagsena. But now, having thrown off the Selucid yoke, Menander, a Bactrian Greek king controlled the entire western region of Sind and Punjab and raided deep into the Ganges Valley right up to the former Mauryan capital of Magadha. The Shungas were also being harassed from the east by King Kharavela of Kalinga. Though he professed to be a follower of the non-violent Jaina religion, Kharavela was as warlike as any, and had once even captured Magadha in about 50 BC. From the ruins of the southern empire of the Mauryans had risen the Satavahanas or the Andhras who, from their capital city of Praisthana (modern Paithana) in the Deccan, constantly attacked the Shungas. Ultimately, in 28 BC, after ruling for almost a century, the Shungas capitulated to their nominal successors, the Kanvas. The Kanvas, in turn, were completely overrun by the Andhras who, around the turn of the century, emerged as the ruling power, at least in southern and central India. Apparently, this turmoil and struggle for temporal power amongst the former vassals of the Mauryans did not disturb the peace of the various Buddhist monk establishments. They continued to build and expand their great stupas and monasteries with unabated vigour.

Hollow Core Stupas on the Krishna and Godavari

The monument that seems to have caught the fancy of the Buddhist builder in the South was the stupa. Numerous examples of this type once dotted the countryside around the deltas of the Krishna and Godavari rivers, the remains of more than ten sites having already been discovered. Unlike those at Sanchi, most of the stupas of the South were not enlargements of existing mounds; rather, the entire hemispherical mound was often built up *ab initio*. While some of the earlier examples such as those at Bhattiprolu and Gudiada were built over a solid core of brick, the builders soon learnt to scientifically economize on the use of their giant-sized kiln-burnt bricks measuring as much as 24 × 18 × 4 in (60 × 45 × 10 cm). Thus, the core of a stupa at Ghantsala *(Fig 6.01)* was formed by an inner circular core of brick walls radiating out much like the spokes of a wheel from the inner hub to the outermost circumference. These cross-walls formed sectoral compartments which were filled with earth and rubble. This economy in construction, however, seems to have been more than amply set off by the lavish use of marble as a casing material for the lower drum, at least, instead of the more austere yellow sandstone of Sanchi. The most sumptuous of this series of southern stupas seems to be the one at Amravati, on the southern banks of the Krishna river.

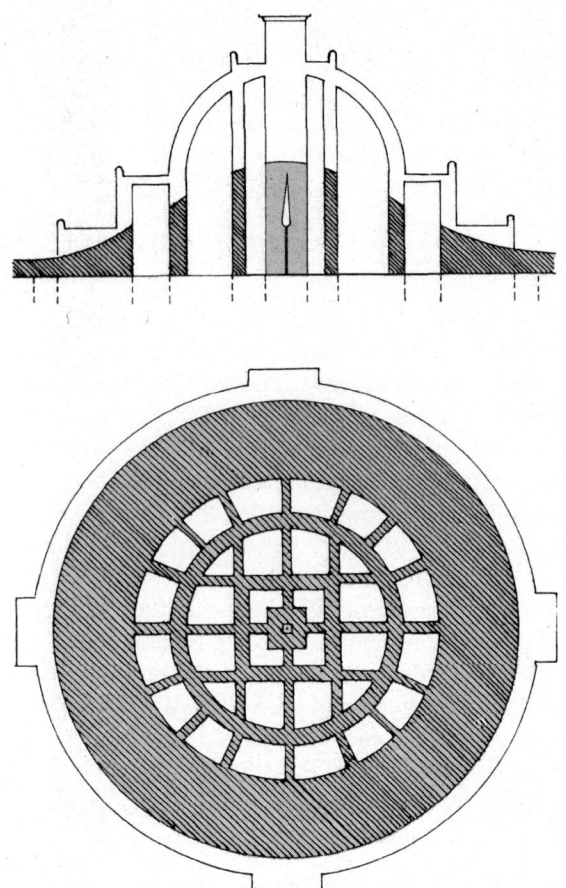

Fig 6.01 Plan and section of the economically structured stupa of Ghantasala, with cross brick masonry walls to contain the earth filling

The Marble Stupa of Amravati

Today, nothing remains of the grand white mound of Amravati except an irregular trench marking its site. The very reasons for its splendour seem to have hastened its desecration; its rich casing of marble made it a target of incessant vandalism. The few surviving marble slabs of its covering are now to be found only in museums in Madras, Calcutta and London. Fortunately, with the Buddhist builders, "it was the practice to include in the carved decorations of the stupa small duplicate copies of itself in bas-relief showing clearly and in full detail exactly what the main structure was like." From such panels *(Fig 6.02)* and the existing remains, it is possible to infer that the Amravati stupa was about 190 ft (57.9 m) in outer circumference, making it the largest stupa ever built in India. It was also the most profusely decorated, both architecturally and sculpturally. The cardinal points of the upper *pradakshinapath* were marked by clusters of free standing pillars or *aryakas* giving an unusual artistic distinction to this southern building type *(Fig 6.03)*. While no toranas seem to have been erected, the entrance to the lower *pradakshinapath* is marked by *stambhas*, which were also planted at regular intervals along the circumambulatory path. In structuring balustrades for the two paths the builders chose solid decorated marble panels for the upper one, while the lower one is a richly sculpted version of the traditional *vedika* of Sanchi.

Centuries ago, the white marble stupa of Amravati must have been a dazzling monument to the glory of the great Lord Buddha who was now beginning to be worshipped as a god and saviour, rather than the mere propounder of a 'middle path' to salvation.

In the meanwhile, Satkarni, the first great Andhra king, became an ardent supporter of the Brahmin faith and proclaimed his religious inclination by performing the Ashwamedha Yagna or horse sacrifice. The Buddhists, meanwhile, in spite of being able to carry out extensive building projects in the plains of the South, and unhindered by persecution by the Andhras, seem to have read the writing on the wall. They were on the lookout for more isolated, secure and hermetic sites to built their monasteries. Their search led them to the so-called ghats, where the Deccan Plateau falls down to the western coast in wild gorges and cliffs. They sensed that the uncertainties and turmoil of temporal power could not possibly penetrate these virtually inaccessible forested hillsides. At the same time, trade routes leading to the river port of Kalyan ran parallel to the foot of the ghats, and mercantile patronage would, therefore, continue to be available.

Fig 6.02 A bas-relief panel from the marble casing of the devastated Amaravati stupa of AD 300

Fig 6.03 Aryakas instead of toranas to mark the cardinal entrance points of the Amaravati stupa

Caves in the Western Ghats

Until the arrival of the Buddhists there had been little building activity of any consequence in this difficult region. During their very first season in the hills, living in thatched huts, the monks must have had to face the fury of the rains which lashed down the hillside almost nonstop for half the year, rains that could wash away an entire village without leaving a sign of it, and rains that would make worship around a stupa in the open an unpleasant task indeed. Under the circumstances, they could have ventured to build halls of worship, like the one at Sanchi. Timber was at hand from the forest. Bricks baked in the plains could, though with great difficulty, be hoisted up the steep hillsides. However, even such a hall, if successfully completed, would be difficult to maintain or preserve under the unrelenting fury of the monsoons. The hills alone, it would appear, could stand up to the ravages of the incessant rains. With their great desire to make the good law of the Buddha outlast time itself, the Buddhist monks, taking a cue perhaps from the Ajivikas of Asoka's time, decided to carve their great sanctuaries out of the living rock of the immovable mountains.

They started with a few simple caverns, presumably to serve as a kind of site office for the 'resident architect' or 'clerk-of-works'. The earliest known of these is at Bhaja. This experimental cutting gave the rock carvers a working knowledge of the amygadaloid and cognate trap formations that made up the vertical bluffs of the ghats. The formations, being of considerable thickness and marked uniformity, were found eminently suitable for the project in hand.

In carving out their halls of worship at Bhaja the Buddhist monks reproduced, as far as possible, exact structural copies of existing *chaitya* halls, but on a much grander scale *(Fig 6.04)*. Their dependence on traditional building forms as models

Fig 6.04 The intricate stone carved details of timber joinery in the caves of Western India, were inspired by timber prototype chaitya halls, constructed with a framework of timber with panels of brick

Fig 6.05 Facade of the 150 BC cave at Bhaja showing timber details incorporated into a stone facade

for their caves should not surprise us. Five hundred years earlier, their Persian 'gurus', in giving a facade to the rock-cut tomb of Darius at Nawsh-e-Rustom, had reproduced *in toto*, the elevation of a conventional built-up Persepolis palace. Closer home, the Ajivikas, in carving their caves at Barabar, had successfully done the same.

Techniques of Cave Architecture

The intention of the monks, then, was to create the familiar apsidal ended *chaitya* hall in the living rock. The rock carvers tended to the technical details. The chosen hillside was cleared of shrubs and other growth. On this roughly vertical rock wall the gable-end elevation of the structural *chaitya* hall was then sketched. Two parallel tunnels were subsequently run to the desired depth and timber wedges driven vertically into the exposed rock at convenient centres. When moistened, these wedges expanded and dislodged large chunks of stone that were removed through the mouth of the cave. Once exposed, the desired surface of the rock was chiselled to final smoothness before being broken further downwards to avoid the tedious and expensive process of later erecting a scaffolding in order to carry out polishing operations. Gradually, with the use of hammers and mere ¼ in (6 mm) chisels as the only tools available, the rock was excavated from the ceiling downwards. The entire volume of the hall—a facsimile of all the timber details of the structural *chaitya* on the inner-surfaces, even down to the transverse rafters in the barrel-vault—was patiently won out of the living rock (*Fig 6.05*).

The sides of the central arched opening in the face of the cliff were now decorated with living rock mock-ups of timber loggias and balconies. Finally, the ever dependable carpenters were called in to cover up the large horseshoe opening with a screen of timber trellis work, imitating the familiar sun window that was traditionally installed in structural *chaitya* halls *(Fig 6.06)*.

The Buddhist clergy must have been immensely pleased with the results, so much so that at various appropriate sites within a radius of about a hundred miles, numerous such *chatiya* halls were excavated. Some of them, at the now famous site of Ajanta (described later) are also a part of this early Buddhist movement.

The concept of recreating original timber structures in stone remained the essence of this movement. Gradually, however, the stone carver began to exploit the unique potential of his own art of rock cutting. Free as he was from the structural restraints of conventional building, he carved out barrel-vaulted hall after hall with great flamboyance, each larger than its predecessor, and more voluminous than any built so far with timber and masonry. Columns that were plain shafts in the earliest examples, were enriched with sculpture and even paintings in later efforts. Under the growing confidence of the stone moulder the *chaitya* arch took on various fanciful and fluid sculptural forms derived from rich Buddhist imagery, particularly that of the lotus flower *(Fig 6.07)*. Timber trellis work on the facade was reduced to the minimum to ensure permanency. The monumental culmination of this peculiar art form that for lack of adequate nomenclature has been termed 'rock-cut architecture', is the well-known mammoth hall at Karli, carved out of the Western Ghats sometime in the first century AD.

Facing page
Fig 6.06 Sketch showing relationship of the rock facades, the intricate carving on it and the long cave cut into the living rock of the hill side

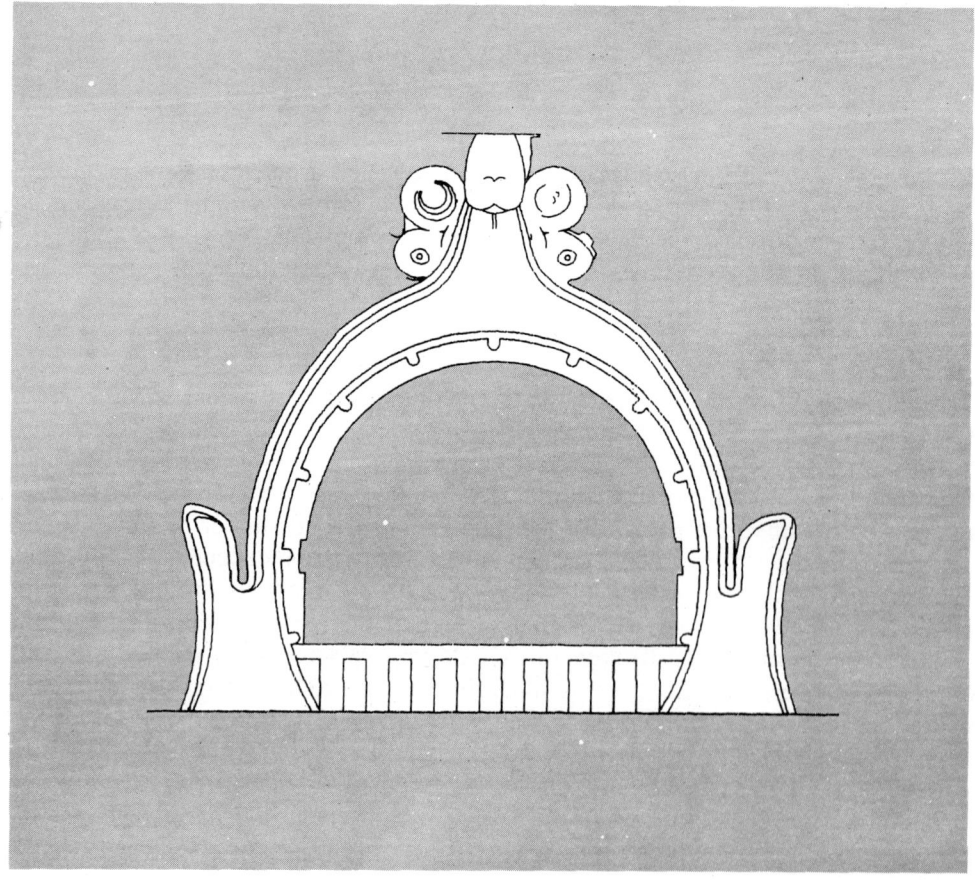

Fig 6.07 The fluid and graceful outlines of bamboo and timber forms were immortalized into window openings for the dark caves

Karli, the Cave Magnificent

In plan, this gigantic hall (45 × 150 ft) (14 × 46 m) is not remarkably different from the others (*Fig 6.08*). The designer of this cave was fully conscious of the potential of visual drama that could confront a visitor entering such a vast sacred hall. There were no mean staircases and ledges in the hillside to reach this *chaitya*. With the debris removed from the cave, a large artificial platform was built over the sloping hillside in front. The pilgrim on the platform is overawed by two enormous 50 ft (15.2 m) high, free-standing masonry columns, crowned by glorious lion sculptures standing like immovable sentinels on either side (*Fig 6.09*). The facade behind is a rock-cut, two-storeyed screen of stately columns. One passes into a double-height ante-room or vestibule and is confronted by the recessed sun window set in a towering horse-shoe archway (*Fig 6.10*). The walls are richly sculptured (*Fig 6.11*), and the flat bare surfaces were at one time painted with colourful scenes from Buddha's legendary life. Through this magnificent facade, one enters the twilight space of the interior and gradually discerns the arrangement of a central apse flanked by massive columns, with two narrower ones on either side (*Fig 6.12*). At the end of the deep cave looms the most sacred object of veneration, a sculpture of the Buddha moulded into the hemispherical profiles of the familiar stupa. The whole concept, execution and achievement of the indefinable space could only be the work of a thaumaturgist. Magic is created by soft effulgent luminosity rather than light that filters in through the outer colonnade and screens and the oriels of the sun window, and falls like a delicate mantle on the stupa situated, seemingly, at the end of a fathomless cavern. It is impossible not to be moved by the solemn majesty of this unique place of worship.

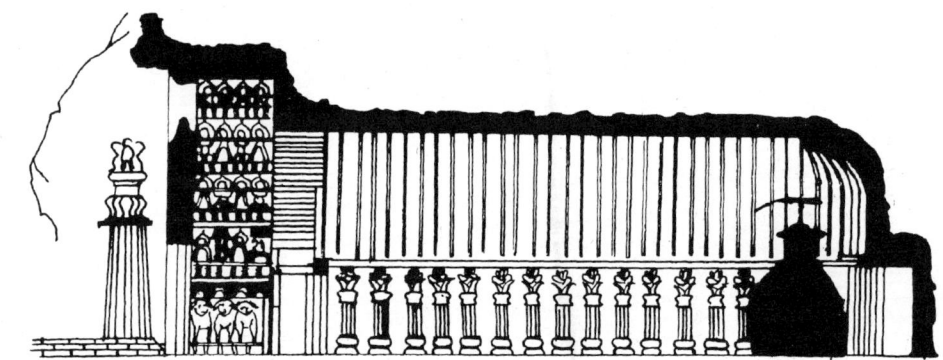

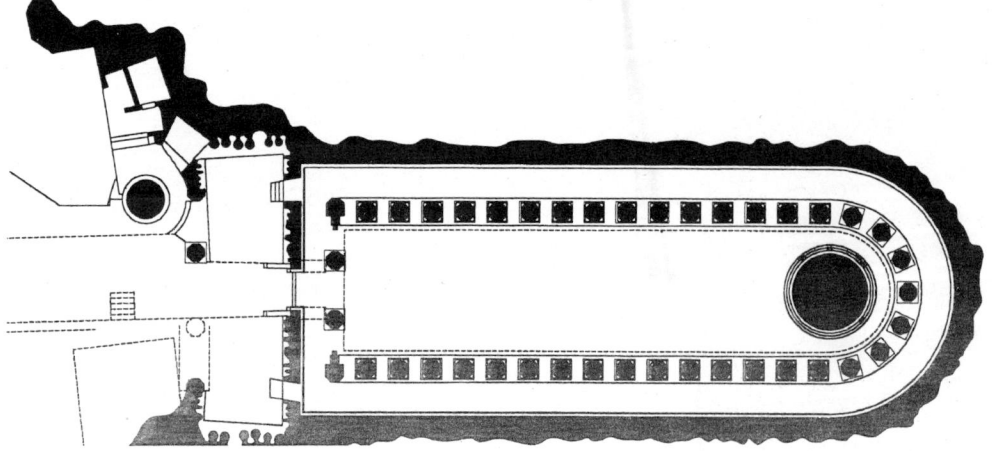

Fig 6.08 Plan and section of the great Karli cave (100 BC) in the Western Ghats of India

➤ *Facing page*
Fig 6.10 The double height anteroom of the Bedsa cave in Western India, bathed in early morning light

Fig 6.09 The Karli entrance was not merely an opening in the cliffside, but included a stately platform adorned with stambhas with lion sculptures

Fig 6.11 One of the richly sculptured walls, Karli Cave

Builders were to return and continue to carve out cave after cave through the centuries at other sites, including the well known ones at Ajanta and Ellora. The latter builders, however, laid greater emphasis on the decorative rather than the spatial aspect. The *chaitya* hall of the first century AD at Karli remained, architecturally at least, an unrivalled wonder of the rock-cutter's art, and an unending source of inspiration to the Indian builder. For, as we shall see, though structural techniques evolved over the centuries, the Indian craftsman remained for ever conscious of the fragile and yet powerful relationship between light, space and sculpture that had endowed the caves of the Buddhists with magical qualities beyond the comprehension of day to day architecture.

Fig 6.12 The magic quality of the space, light and sculpture in the cave architecture of Buddhist India is something that has to be felt rather than merely seen. At Karli it is impossible not be moved by the "solemn majesty" of this unique place of worship

CHAPTER 7

The Greeks, Kushans and the 'Larger Vehicle'

While Buddhist monks in western India had sought refuge from the vicissitudes of unending feudal wars in secluded inaccessible sites, those in the North continued to play an active role in the rapidly changing political situation of the trans-Indus Gandhara region. The Greek king, Menander, who had earlier made raids into central India, gained firm control over the region of the Punjab and Sindh and ruled over it from his capital at Sagala (modern Sialkot).

The city of Sagala has been described by a contemporary chronicler in glowing terms as "laid out by wise architects, with many and various strong towers and ramparts, splendid with hundreds of thousands of magnificent mansions and bazaars which face all quarters of the sky." Today, there is no trace of the towers, the mansions, or bazaars. During its heyday, the Buddhist monk, Nagasena, visited the capital with his disciples "lighting up the city with their yellow robes like lamps." Nagasena had acquired reputation and fame enough to prompt King Menander to go out and receive him personally, accompanied by his retinue of royal courtiers. There followed a catechismal discussion with Menander—the famous 'Questions of Milinda'—that is said to have resulted in the king being converted to Buddhism.

In order to win such disciples, preachers of Buddhism were beginning to enlarge the scope of the simple doctrine preached by its founder. Buddha was accepted by Nagasena as 'god' and 'saviour' and his graven image was now worshipped as a deity (*Fig 7.01*). This interpretation of Buddhism became known as the Mahayana or 'larger vehicle', while Buddha's original doctrine was designated as Hinayana or 'smaller vehicle'.

Fusion of Roman, Greek and Indian Styles

With the assured royal patronage of Menander, Buddhist monks settled down to the task of erecting stupas and monasteries. They were in a hurry to capitalize on their good fortune, and unhampered as they were by orthodox Hindu insularity, they had little qualms in using the talents of immigrant craftsmen who were better acquainted with Roman and Greek, rather than purely Indian traditions.

The edifice, generally known as 'the shrine of the double-headed eagle' at Sirkap (near Taxila), is a square platform faced with local Kanjur stone, adorned with a grid of pilasters of the Roman composite style (*Fig 7.02*). Over this once stood an elongated version of the Sanchi type stupa that has since vanished. The spaces between the pilasters are filled with reliefs of Roman classical pedimented aedicules, as well as the indigenous *torana*, and a *chaitya* arch. This expedient mixing of Romano-Greek and Indian motifs is the embryo of the architectural style popularly termed Gandhara, after the region in which it was born.

Fig 7.01 A carven image of Buddha in the Gandhara style marking the acceptance of Buddha as "god and saviour".

The support of Indo-Greek kings to Buddhism was unfortunately short-lived. They had ruled barely for a hundred odd years when a series of chain reactions beyond the frontiers of the country of their adoption brought the armies of the Scythian tribes to the borders of India in 80 BC.

Fig 7.02 Buddha resting within a typical Roman Greco column, capital. The fusion of Indian, Roman and Greek traditions is the hall mark of the Gandhara style

The Scythians and Kushanas

The Chinese emperor Chi Hsuang Ti, who was responsible for completing the construction of the Great Wall of China, effectively closed off the pastures of China to nomadic tribes of the Yeu-Chi. The Yeu-Chi were forced to press on to the region west of the Aral Sea, original homeland of the Scythians. The displaced Scythians in turn overran Parthia (the northern regions of modern Persia) and Bactria, and swept into India probably through the Bolan Pass near Quetta. They had little difficulty in defeating Hippostratus, the last known Greek king.

The brief rule of the Scythians, or Shakas as they came to be known, did not leave any impression on the Buddhist architecture of the times. The classically inspired Fire Temple of about this period at Jandial near Taxila (*Fig 7.03*) is a curious mixture of a Greek motivated architectural shell for the performance of Achmenid Persian religious rituals. In plan, it is a classic Greek peripheral temple in antis, with columns of the Ionic order. There is, however, no imagery, and the platform at the back of the temple suggests the presence of a wooden fire tower dedicated to Mazda, a god of the Zoroastrian pantheon, probably worshipped by the Scythians and Parthians.

The rule of the Shakas in this region was put to an end by their old tormentors, the Yeu-Chi. Led by Kujala Kadphises, they descended through the legendary Khyber Pass in about AD 50. The Shakas retreated to seek their fortune deeper into western India where they were ultimately absorbed into the Hindu fold. The invaders in the north-west established themselves as the well-known Kushan dynasty of Indian history.

The greatest of the Kushan kings was Kanishka, who ascended the throne in AD 78, and ruled over a kingdom that extended in the south up to Sanchi, in the east to modern Varanasi, and in the north-west included the entire Gandhara region. The two great foci of this empire were the cities of Purushapura (modern Peshawar), and the famous Mathura on the banks of the Yamuna. The favours of an emperor as powerful as Kanishka were eagerly sought by both the Brahmins and the Buddhists.

Fig 7.03 The 200 BC fire temple of Jandial is a curious mixture of a Greek architectural shell amended for the performances of Zoroastrian rituals. 1 Naos, 2 Pronaos, 3 Peristyle

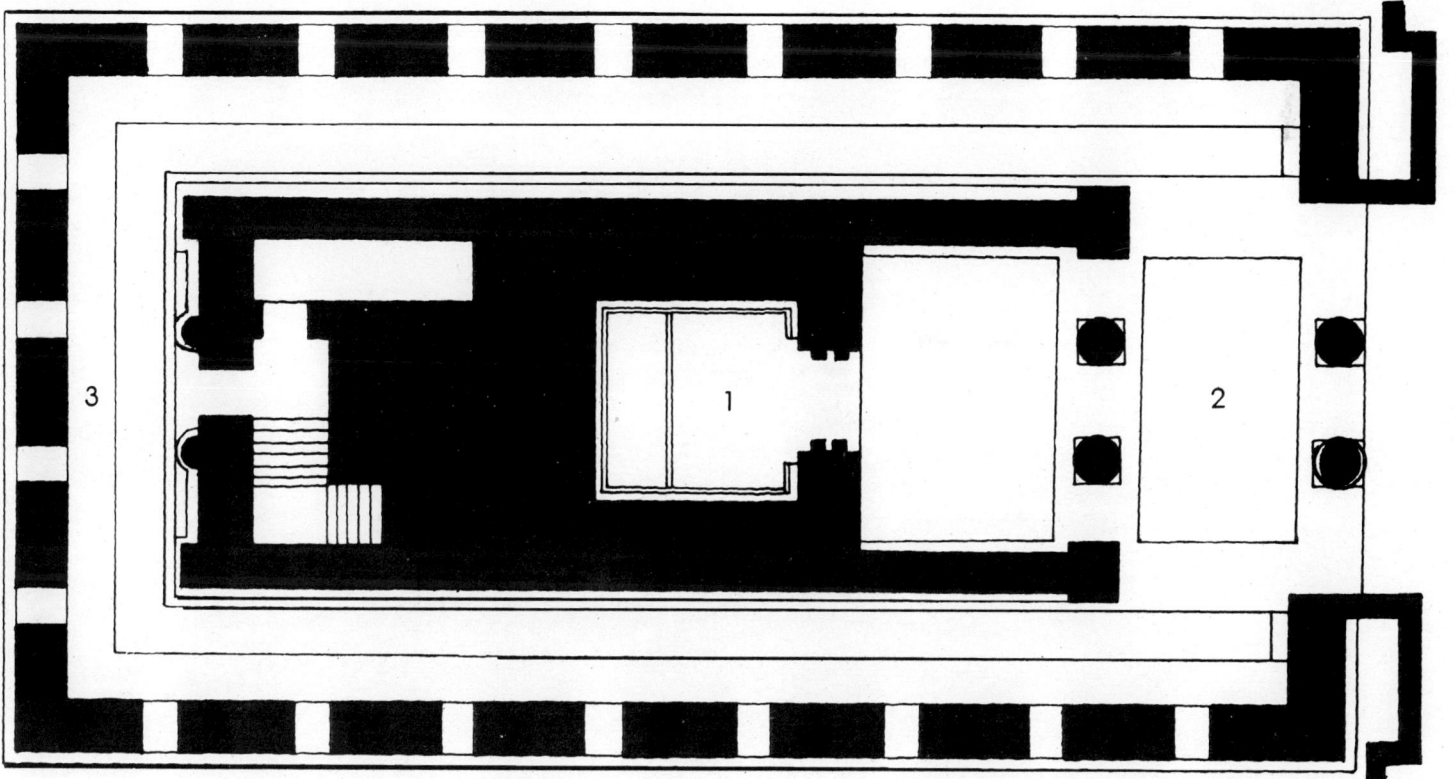

Mahayana Buddhism with its direct, more simple approach had a stronger appeal for this great warrior king than the involved metaphysical theories of the Brahmins, overlaid as they were by innumerable and vexatious caste restrictions.

Asvaghosa, himself a Brahmin convert, was ultimately able to persuade the emperor to adopt Buddhism as his state creed. In Kanishka, Mahayana Buddhism had won a patron of as great significance as Asoka was to Buddhism in its early days. It was at the behest of Kanishka that the fourth and last Great Council of the Buddhists was held at Kundalavana monastery in Kashmir, a perennial favourite of these central Asian kings. Much like the later Moghuls, the Kushan kings often retired to the sylvan valley to escape the dreaded heat and dust of the Indian plains.

Prosperity under Kanishka

Kanishka's empire was, if anything, richer than Asoka's. In the wake of the unending foreign invasions of the Greeks, Parthians, Scythians and the Kushanas, trade routes, both by land and sea, had opened up to the west. The Kushanas were on excellent terms with the Romans, the borders of both empires at one time being just 600 miles apart. An immense volume of condiments, spices, unguents and silk left from the ports of later day Gujarat and Sindh, in exchange for Roman sureis, Greek wines and choice girls for the royal harems.

Such lucrative overseas trade gave rise to a prosperous commercial class. The Buddhists were fortunate in enjoying dual patronage; that of the state, and of their ever dependable allies, the mercantile community. They were busy building once again. Over a period of a couple of hundred years, a number of richly patterned stucco painted and even gilded shrines and monasteries of the Buddhists dotted the otherwise "bare crests and featureless slopes of the Gandhara countryside."

Fig 7.04 The Takht-e-Bahi monastery near modern Peshawar, nestling between bare hillsides, typical of the Gandhara countryside

Monastery of Takht-E-Bahi

One of the more impressive and schematically planned of these settlements was built at Takht-e-Bahi (modern Peshawar) (*Figs 7.04, 7.05*). A 50 ft (15.2 m) high elongated boat keel profiled stupa rested on a 20 ft (6 m) wide square platform, set within a 45 ft (13.7 m) by 55 ft (16.7 m) quadrangle along the periphery of which were subsidiary chapels housing the Bodhisattavas of the Mahayana creed. The roofs of the cells of the surrounding chapels were, alternately, a dome derived from the shape of the rural beehive hut, and a barrel-vault derived from the roof of the *chaitya*. Constructed in stone, the roof shapes were built up by means of corbelling, rather than true arching. The masonry, with various sizes of stone arranged in a diaper fashion, was covered with a thick layer of lime plaster relieved with stucco decorations.

The central axis of the court of the stupa leads to the familiar monastery, which is a series of cells around its own quadrangle (*Fig 7.06*). In between the stupa court and the monastery court is an open terrace adorned with small chapels and votive stupas. Around this well-knit plan of the central area, are other haphazardly situated ancillary structures, such as an assembly hall, refractory and vestment chambers.

Fig 7.05 The Takht-e-Bahi monastery, another view

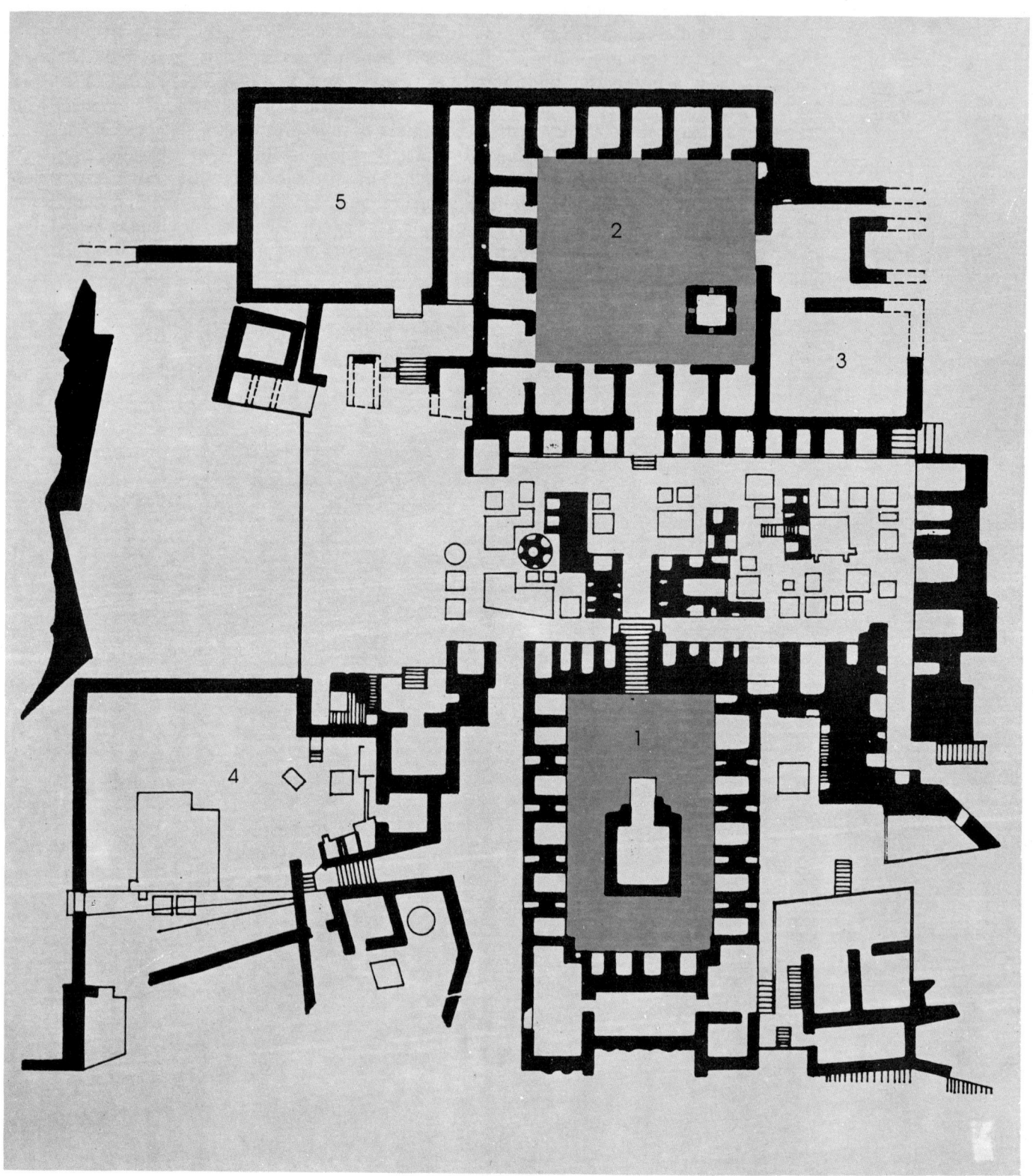

*Fig 7.06 1 Court of the stupa,
2 Monasteries, 3 Assembly hall,
4 Refractory and 5 Vestment
chambers of Takht-e-Bahi
monastery*

The Ali Masjid Stupa

The better preserved Ali Masjid Stupa in the Khyber Pass, gives a clearer idea of the decorative scheme of the Gandhara stupas. The stucco frieze applied to the base of the drum is a series of *chaitya* arches, supported on Corinthian pilasters, each niche being filled with a Grecian version of the different aspects of the deified Buddha of Mahayana.

The stupa, as may be seen, was gradually being transformed from the domical into an elongated and vertical structure (*Fig 7.07*). Apart from the almost primitive desire to make a religious memorial ascend towards the skies, this transformation may well have been influenced by the environment. The traditional white hemispherical mound that was a sufficiently striking element in the plains, would have merged inconspicuously into the slopes of the bare hillsides. However, the bell-shaped stupa stood out against the background of a rolling countryside. The vertical umbrella and *harmika* at the top gradually came to dominate the entire composition, often becoming taller than the brick and plaster mound over which it was mounted. As can be seen, this was not an altogether pleasing or logical architectural imposition (*Fig 7.08*). Obviously, the aim of the builders was to create taller buildings, without too much attention to the aesthetics of proportion.

Fig 7.07 The Buddhist builder was encouraged to elongate his stupa into a vertical monument so that it would be distinguishable from the gently rolling Gandhara countryside

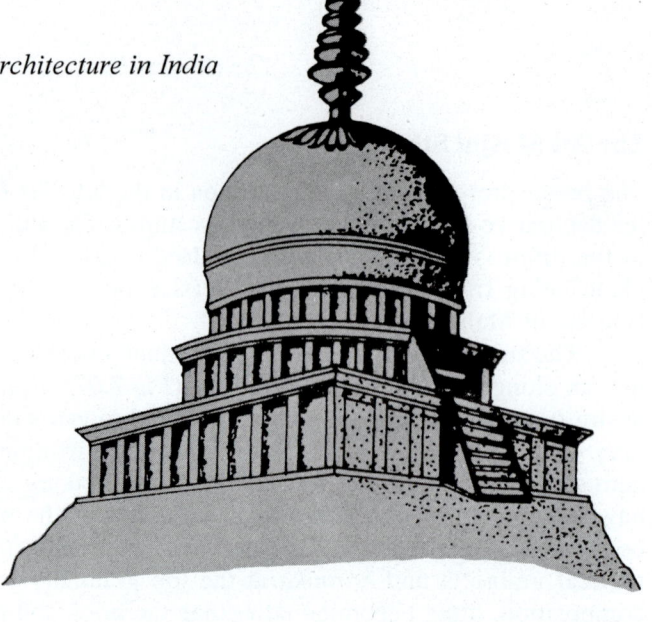

Fig 7.08 *Platforms in the Greek tradition became the base over which stupas with an elongated umbrella were planted. This accent on height culminated in the design of the famous Chinese pagodas*

The Prototype of the Pagoda

The passion for verticality culminated in a tower said to have been built by Kanishka in the first century near modern Peshawar. All traces of it, except the ruins of its base, have vanished. It was a veritable Buddhist wonder of the world. As Shun Yat Sen, a Chinese pilgrim of the sixth century testifies, the timber tower of this structure was 628 ft (190.7 m) high consisting of 13 receding storeys. The whole was surmounted by an iron pinnacle which was adorned with copper gilt umbrellas. It rested on a massive 285 × 285 ft (86 × 86 m) platform of brick, and was ultimately destroyed by an attack of lightning, sometime in the seventh century. Chinese pilgrims, facilitated by the contact between the Kushana and Chinese emperors, had before its collapse carried descriptions of the marvellous edifice to their homeland where the design took root, and under the genius of the Chinese carpenters, flowered into the familiar pagodas of China.

Mahayana School of Mathura

Mathura, the more centrally situated city of the Kushan empire was also humming with Buddhist, and even Jain architectural activity. The building style here was more indigenous, inspired as it was by the art of Sanchi and Bharhut. The spiritual content, however, was of the Mahayana school. Fa Hien, another contemporary Chinese pilgrim to the holy land of Buddhism, saw over a score of monasteries in Mathura inhabited by more than 300 monks. A number of stupas and towers were also erected here. These, unfortunately, were so thoroughly demolished by the iconoclastic zeal of later invaders, that not a single complete building of this period has survived.

Decline of the Kushans and the Andhras

Kanishka's long and prosperous reign came to an end in AD 162 when he died at war with the Chinese somewhere in central Asia. He was succeeded by his sons Vanishka and Havishka who remained patrons of Buddhism. But the Kushan empire was being threatened from the west by the newly rising and powerful Persian dynasty of the Sassanians. The last Kushan king, who had even taken on the Indian name of Vasudeva, was pushed east of the Indus, and ruled for a short period over a small kingdom around Mathura.

With the end of the Kushana dynasty in AD 220, the trans-Indus region had once again passed under Persian dominance and was cut off, for some time to come, from the mainstream of Indian history.

CHAPTER 8

The 'Sanskrit' Brahminism
of the Guptas

Looking back southwards, we find that the power of the once great Andhras was also wilting under the pressure of various burgeoning dynasties of the Deccan. When the major powers of the north-west and south-central India collapsed simultaneously in about AD 220, the two empires were dismembered into a number of provincial kingdoms whose rulers had been chafing under the centralized power of the Kushans and Andhras. The Kushan empire gave way to the Yaudehas and Malavas in Rajasthan, and to the Nagas, Uccalpas and Lichhavis in eastern and central India. Suzerainty over the former Andhra country was now shared by the Vakatakas, Chalukayans and Sakas, while the southern tip of the country passed into the control of the Pandyas. The central-southern area fell to the Gangas and Kadambas, the Pallavas holding the south-eastern coast.

Prosperous All-absorbing Hinduism

A number of dynasties, like those of the Sakas, Yaudehas, Vakatakas, and some believe even the Pallavas, were all descendants of various tribes that had been successively pouring into India through its north-west defences only over the last couple of hundred of years. Like the Sakas, they had been gradually pushed deeper into India by subsequent invaders and ultimately carved out a niche for themselves in the complex all-absorbing structure of Hindu society. The country of their adoption was at this time riding the crest of an economic resurgence. It had accumulated tremendous wealth through its trade with the contemporary Romans, Arabs, and even the south-east nations of Java, Bali and Sumatra. The rulers of these essentially immigrant dynasties were therefore keen to identify themselves completely with India and to share fully in the material wealth and prosperity of the country.

The feudal kings, following the lead of Kanishka, were beginning to adopt Sanskrit, spoken by the Aryan Brahmin, as their official court language. Since Buddhism had already acquired the status of a world, rather than an Indian religion, Brahmins were quick to present themselves as the only bonafide torch-bearers of a truly indigenous culture to appeal to the prurient interests of the settled tribes and the chauvinism of the ancient Indians. Moreover, in this time of prosperity and optimism, the Brahmins' attitude of "God is great, glorious, gracious! God's work are wonderful. God is love! He is so overwhelmingly adorable that His beings can but sing hymns in His glorification," was a more appealing religious attitude to adopt than the comparatively pessimistic outlook and austere message of Buddhism.

Until recently, the Tamil-speaking kingdoms of the south had been at constant and ferocious war with each other, even indulging in acts of cannibalism. They were, however, now organizing and uniting themselves into a maritime power to acquire wealth through overseas trade. With growing prosperity the Southerners too, were attracted by the cultured and courtly ways of the north and began to adopt what may be called Indo-Aryan attitudes. The seeds of a nation-wide revival of Brahminism, popularly referred to as 'Hinduism', were beginning to take root.

Chandragupta

Initially, the Vakatakas tried to take advantage of this undercurrent of national unity, and weld the small kingdoms together into a Hindu empire. The task was, however, ultimately accomplished by Chandragupta, a feudal Raja of ancient Magadha. He emerged out of obscurity, by capitalizing on his marriage to a princess of the distinguished and ancient clan of the Lichhavis in AD 320. In his short reign of just a decade he became lord of Bihar and parts of Bengal and the Ganges plain, founding the dynasty of the Imperial Guptas. This dynasty heralded what has popularly come to be known as the 'Golden Age of India.' Chandragupta's son Samudragupta, continued to capitalize on his Lichhavi blood, and in the course of the next 50 years led his armies to distant parts of India. He conquered most of northern India and made the rulers of Assam, Nepal, Kanauj and Punjab his tributaries. The Sakas of western India who had survived the invasions of Samudragupta were finally subdued by his son Chandragupta II the Great, who raided even as far afield as Afghanistan. He even brought the powerful Vakatakas and Kadambas into his fold through matrimonial alliances.

By the beginning of the fifth century AD the empire of the Guptas was beginning to show shades of the Mauryas, shades only, for the Guptas, in spite of their far flung conquests, found it difficult to install centralized authority over proliferating feudalism, like Asoka had been able to. Their cultural conquest of India, however, was more effective, based as it was, on the secular and liberal traditions of an aristocratic upper class—a class that was fond of refined manners, poetry, art and conversation in high-flown Sanskrit only.

The Buddhist Pressure

Direct state patronage of Buddhism may not have been part of the liberal tradition of the times. Nevertheless, the Buddhists were never the victims of a witch hunt by the ruling class, except when the orders of the monks invited imperial retribution by indulging in the craft of politics. The Sanghas or the orders of Buddhist monks, had in fact acquired large tracts of land. These were a permanent source of income for the building of their great monasteries, and now, even educational establishments. The nuclei of what was to expand into the great university of Nalanda and the monastic complexes of Sarnath and Gaya, were in fact established during this period.

Brick and Timber Chaityas

In terms of wayside shrines, however, the Buddhists were still content with building modest *chaitya* halls. Most of these have vanished, but the surviving one as at Ter and Chezarla in the Deccan (*Fig 8.01*), are fair examples, though converted now to Brahmin worship. These *chaityas* were of modest size, the better preserved one, Ter, being only 22 × 35 ft (6.7 × 10.6 m), in plan (*Fig 8.02*) and 22 ft (6.7 m) high at its apex. The vaulted roof of the *chaityas* was constructed by oversailing horizontal courses of brick masonry. A thick layer of plaster inside and out was used to create the curvilinear profile of a vault.

➤ *Facing page*
Fig 8.01 The Ter chaitya hall with later modifications to convert it into a place of Hindu worship (AD 500)

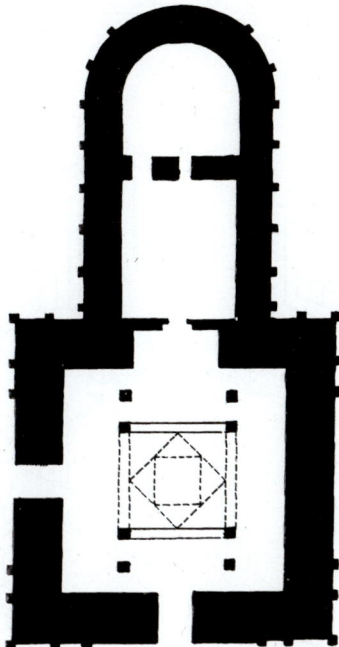

➤

*Fig 8.02 Ter is the only
example of an extant brick
Buddhist chaitya hall that is
now a Hindu temple, the
original duly plastered and
painted many times over*

Pulsating Frescoes of Ajanta

The Mahayana Buddhists, however, lavished their wealth and new artistic activity on the proven media of rock-cut cave architecture. In AD 450 they returned again to the cliffs of Ajanta, abandoned some 200 years earlier by the Hinayana Buddhists, and set about the task of making numerous additions to an already sacred site. The architectural concept of the chaitya halls of the Mahayana followers differs little from the earlier efforts except that images of Buddha became part of the iconographic scheme. The *viharas*, however, emerged now as a combination of a place of residence as well as worship (*Figs 8.03a, b*). Vast square halls were surrounded by colonnades and monks cells. They were entered through verandahs carved out in the face of the cliff and ended at the back, in mini-chapels full of huge images of Buddha and the Bodhisattavas (*Figs 8.04, 8.05*).

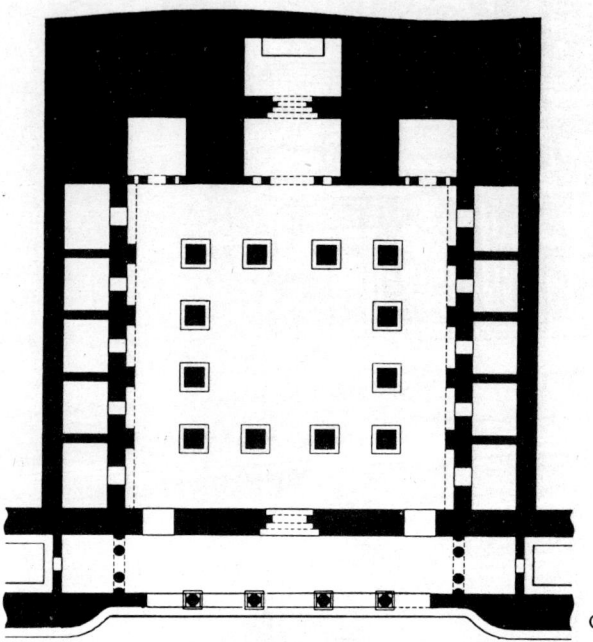

a

Fig 8.03 A group of cells around a central hall of worship; a typical vihara at Ajanta. (a) Plan, (b) Isometric view

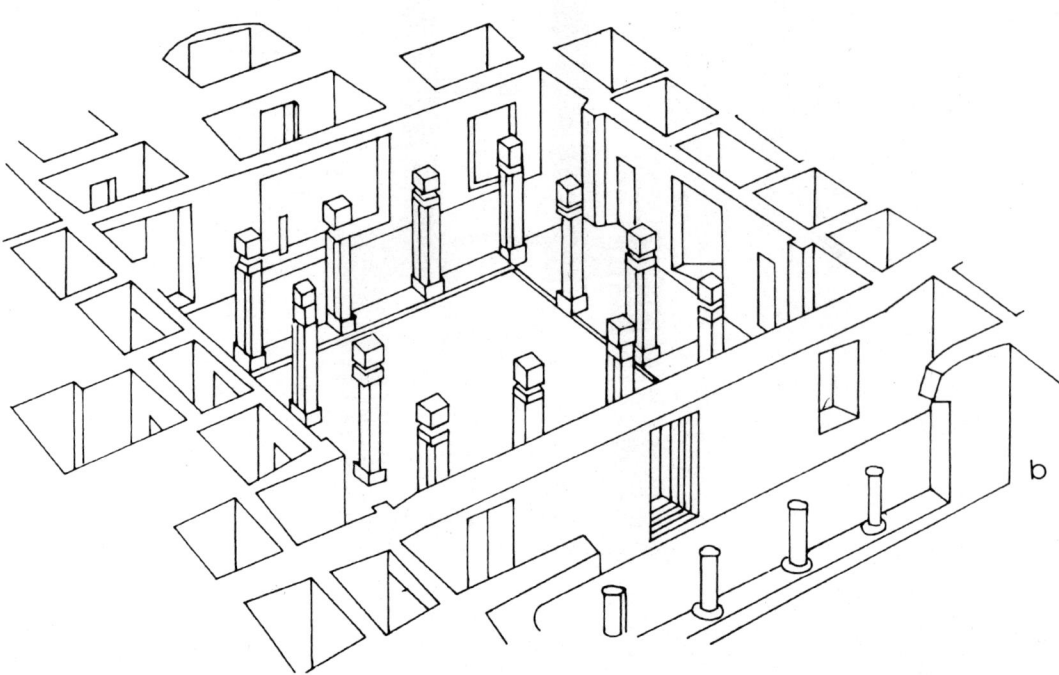

b

Fig 8.04 Facade of Buddhist cell at Ajanta

Apart from carving the traditional sun windows over the main entrances and fascimiles of bamboo gable arches over the doors to the cells, the artists of the Gupta period now proceeded to adorn the walls of the *viharas* with the world famous frescoes of Ajanta. To achieve the effect their first step was to cover the rough surface of the exposed stone with a layer of clay and cow-dung mixed with chopped straw and rice husk or animal hair, finished with a fine coat of gypsum. The flat surface thus created was ready to paint on. The basic design was outlined in red ochre, and coated with transparent monochrome. Then with colours made from natural dyes they depicted elaborate scenes of kings and queens, soldiers and courtesans, monks and merchants, travellers on horse back or in palanquins, in palaces and houses, gardens and forests, set among the flora and fauna of western India (*Fig 8.06*). These frescoes, like early Buddhist sculptures, are yet another example of how the Indian artist with his sharp eye for detail filled the gap created by the traditional disdain for history writing, in our knowledge of an ancient Indian society.

Fig 8.05 *Plan of the group of caves at Ajanta with inset photo of the horseshoe cliffside into which the caves are carved*

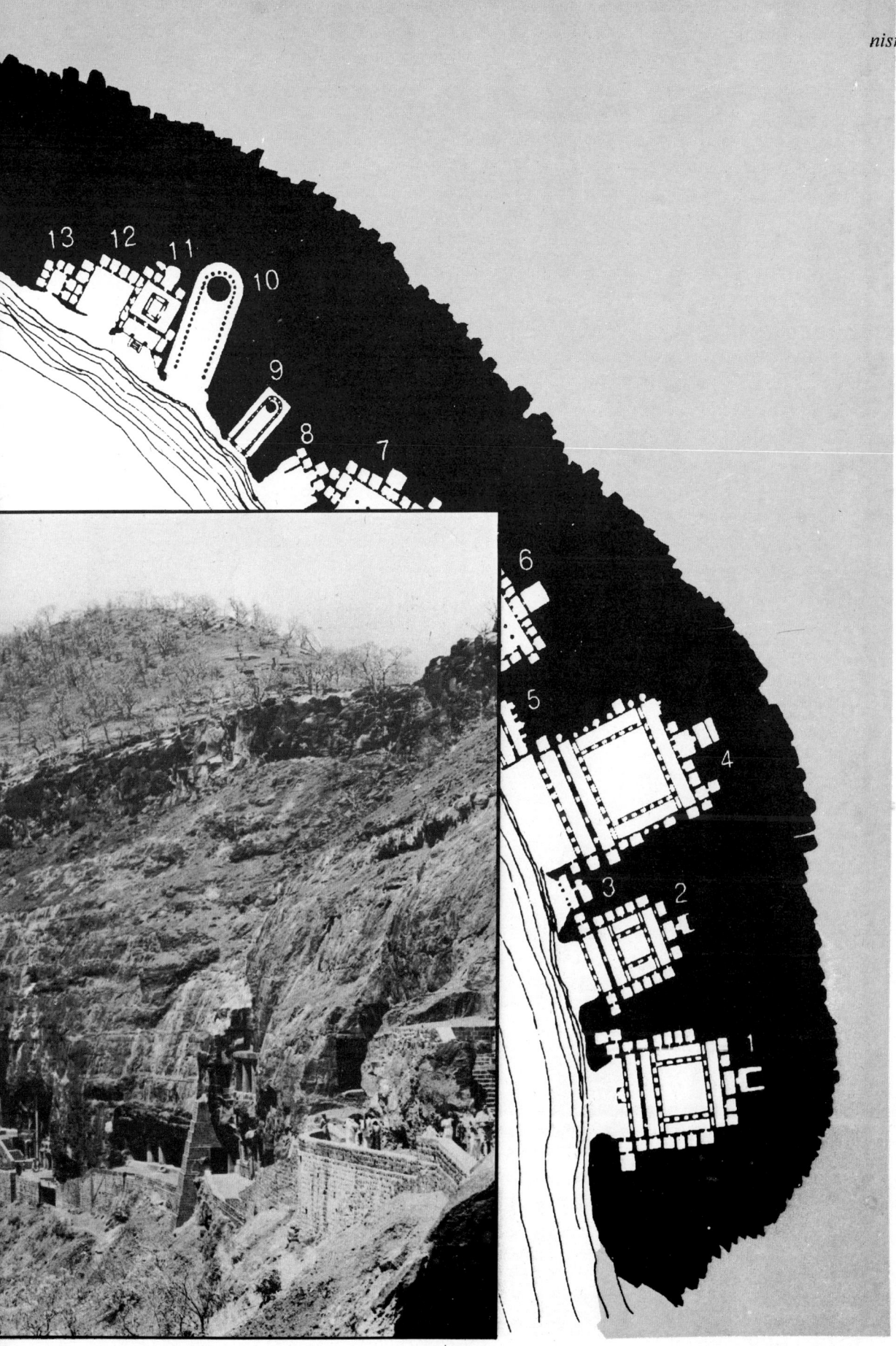

◄ Facing page
Fig 8.06 In painting the frescoes at Ajanta, the Buddhist artist with his sharp eye for detail, provided living images of social life in ancient India

The ceiling of the central *vihara* hall, on the other hand was ornamented by figurative and geometric designs, consciously unrelated to the surrounding architectural elements, to create an effect such as would permeate the blossoming gardens in the open-to-sky quadrangle of traditional structural monasteries.

To facilitate the artist in executing his ambitious works of art, the darkness of the caves was lit up by reflecting the rays of the sun from large metal mirrors installed outside. Under circumstances more trying than those faced by Michalengelo a thousand years later, the artists of Ajanta produced a complete and colourful array of the everyday life of the leisurely and the wealthy. The caves of the Buddhists had come a long way from the sombre and dignified awe-inspiring halls of Karli to the vibrant and pulsating *viharas* of Ajanta.

At the time that Buddhist architecture was reaching the acme of its art, Hindu revival, thriving under the stable rule of Chandragupta II, was seeking an architectural representation for its religious beliefs.

Early Hindu Shrines

The ritual of the Aryan Hindu had till now involved the propitiation of abstract forms of natural forces. These forces being assumed to be omnipresent, the building of permanent earthly abodes for them had not been found necessary. The only edifice required was an altar, roofed, if at all, with timber posts and beams covered with matted reeds. However, like Mahayana, which in bowing to popular appeal had given in to the worship of images of Buddha, an anthropomorphic conception of a deity in the form of a statue, gradually became a part of Hindu worship too.

The earliest Brahminical or Hindu shrine was nothing more than a cell to house an image. Other rituals were still conducted in the open air. The oldest shrines of the Hindu religion, true to the national genius, are once again a group of caves, carved out of a mountain side at Udaygiri, near ancient Sanchi (*Fig 8.07*). The interior is a plain rectangular room or *garbhagriha* (literally the womb-house), in which the image was installed. The entrance is through a small portico, which was a faithful representation in stone of a familiar post and lintel timber model oft-depicted in the paintings of Ajanta.

Fig 8.07 A cave and attached portico at Udayagiri. True to Indian tradition, the earliest place of Hindu worship was a dark cubic space carved out of living rock

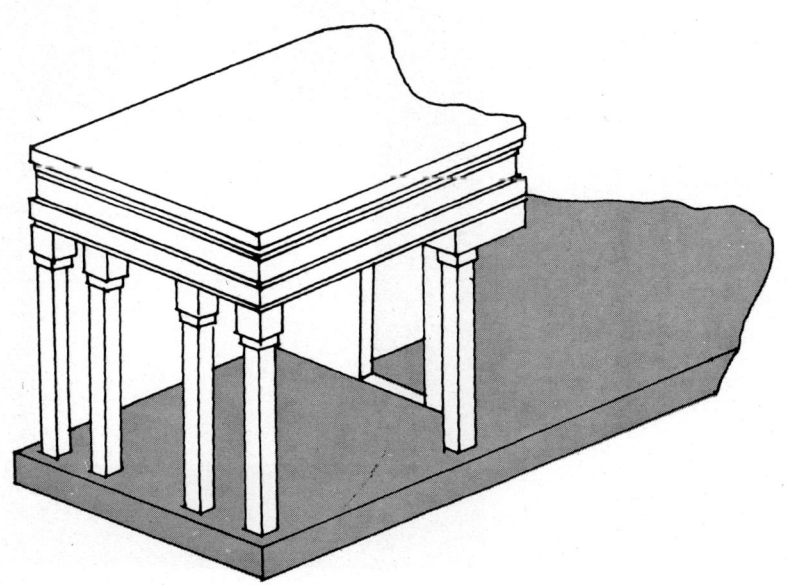

Temples at Tigawa and Sanchi

The earliest extant structural Hindu temples are at Sanchi and at Tigawa near Jabalpur. Though their patrons decided to build temples by the more conventional structive methods of masonry, the only model of Hindu sanctums known to the designers were the grottos of Udaygiri. The wheel now seems to have come full circle. We find the structural masonry builder taking his cue from the art of the rock cutter, who centuries ago had modelled his own craft on free standing timber edifices of the Aryan wood worker.

The Sanchi temple (*Fig 8.08*) of the early fifth century repeats *in toto* the plan of the Udaygiri shrines. The cell, housing the image, is a dark hollow cube. The meticulously jointed stone block walls of the cell support a roof of flat stone slabs. Derived as it was from the concept of a cave, the craftsmen had not yet developed an aesthetic system to embellish the plain masonry of the outer surface. The pillars of the entrance portico attached to the only opening in the *garbhagriha* are stunted versions of the familiar free-standing Asoka columns, complete with the inverted bell capital and imagery of animal motifs, the only kind of vertical shaft known to the Indian builder.

Using stone as a building material for their religious edifices for the first time, the craftsmen unsure of its structural potential, were unwilling to take any risks. Even the roof of the small 12 ft (3.6 m) wide portico is supported on as many as four columns. To allow for a reasonable entrance in the middle, though, the columns are erected in pairs, situated at either end of the longer length.

Fig 8.08 The Gupta temple at Sanchi, inspired as it was by Udayagiri, is a windowless cubic volume with an attached portico

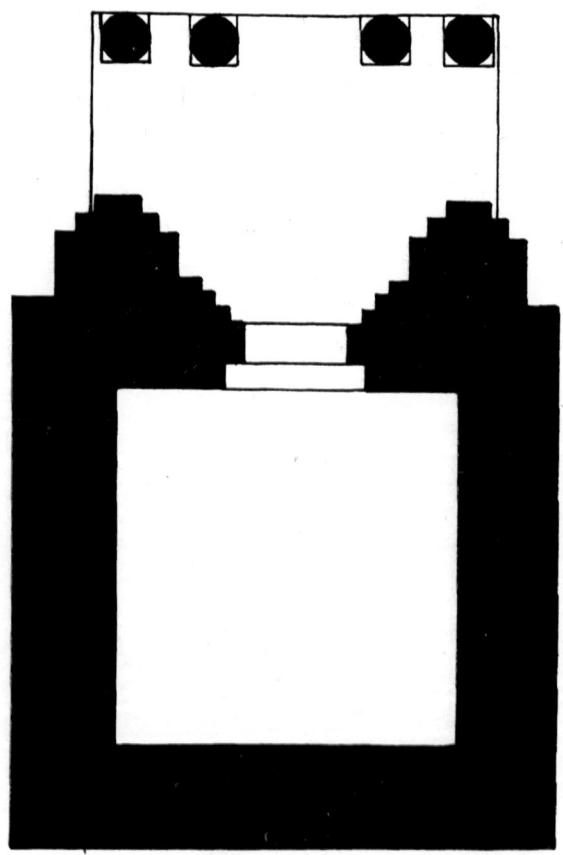

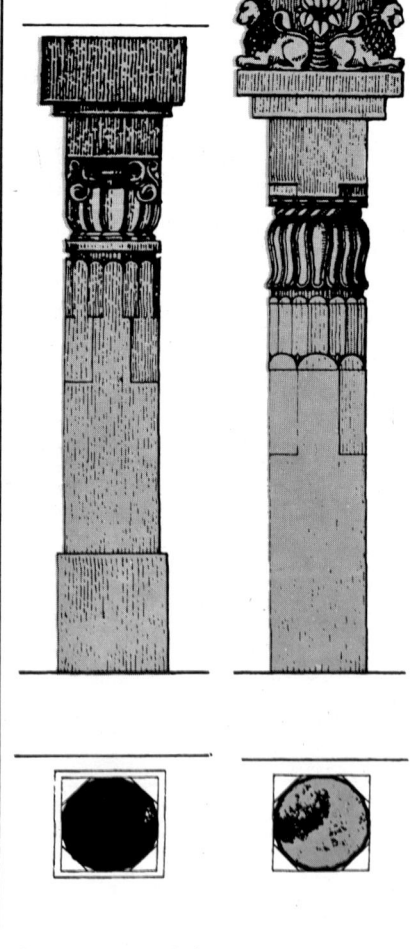

UDAYGIRI CAVE 1   SANCHI

In comparison to this rather subdued effort at Sanchi, the temple at Tigawa, though built on a similar plan, has a refreshing new flourish about it. This is apparent in the novel design of the portico columns (*Fig 8.09*). These are no longer the conventional octagonal shaft of Asoka's period. Rather, square in plan, they are boldly ornamented with robust sculptures of gods and goddesses of the ever-growing Hindu pantheon in flamboyant postures.

The design of the capital, too, heralds the new Gupta imagery of the *purana kalasa*, marking the end of the period of stereotyped inverted bells used consistently for almost 600 years since Asoka's time. The spirit of the liberal Gupta empire was beginning to percolate down to its craftsmen too. As if drunk with this spirit, they proceeded to create new architectonic symbols of the new and resurgent religion of the Brahmins.

Fig 8.09 The portico of the Tigawa temple with the typical Gupta column along with a comparative study of contemporary columns

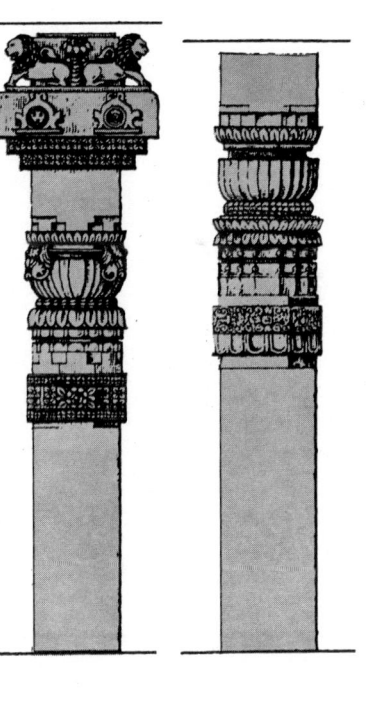

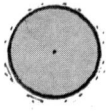

TIGWA   ERAN

CHAPTER 9

The Chalukyas and the Guptas

The empire that Chandragupta and Samudragupta had won through diplomacy and war over a period of 100 years was finally bequeathed to Kumaragupta I. During the 40 years of his peaceful rule, the Guptas reached the zenith of their power and splendour. Chandragupta II had earlier at Ujjain, rallied around him the famous 'Nine Gems', the greatest intellectual lights of his time, setting the characteristic Gupta lifestyle—"a consciously refined syncretism of method, learned training and self-control." Under Kumaragupta, this style acquired an authority over other associate courts, not unlike that wielded by Versailles, in the seventeenth and eighteenth centuries, over other European courts.

Hindu Art Pervades the South

Some sense of this artistic authority, as we shall see, seems to have been felt even in the south. A group of craftsmen were commissioned by their Chalukyan overlords in about AD 600 to build shrines for the Aryan gods, whom they had recently started worshipping, at their capital city of Aihole (in modern Karnataka state). Probably immigrants from the site of Ajanta just over 300 miles north, the craftsmen were familiar with the paraphernalia of Buddhist ritual, and well versed in the high traditions of Gupta sculpture. However, they had neither a visual idea nor any experience in the building of free-standing monuments, leave aside a specific concept of a Hindu temple. Their client, the Hindu priest, too, was as much at sea. He had only the vaguest ideas of what his temple should be—"an impressive structure where a graven image of God could be installed, consecrated and worshipped." True to instinct, the pioneer builders picked on the most prestigious rural timber structure, the 'santhagar,' or assembly hall of the village council, as their model (*Fig 9.01*).

Fig 9.01 The "santhagara" or assembly hall of the village council

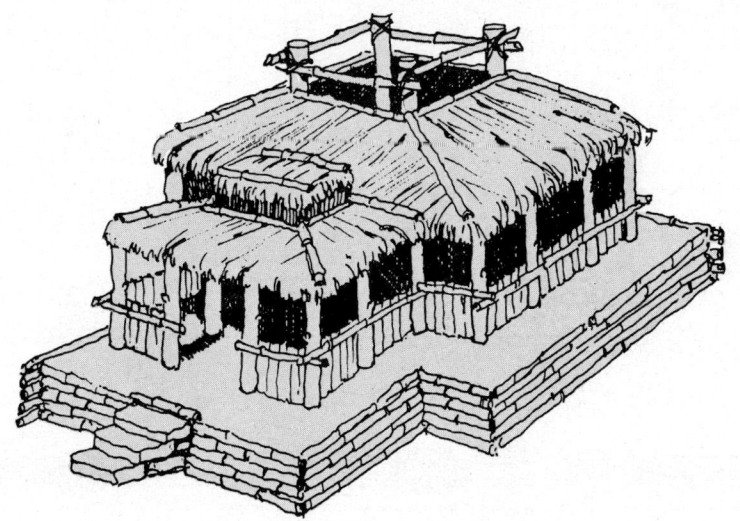

The Lad Khan at Aihole

With a few modifications they set about the task of adapting its timber structural form to the permanency of stone masonry, and its village council plan to one fit for the worship of gods. The result of their innocent endeavour is a temple, which for some curious reason is popularly called the Lad Khan (*Fig 9.02*). In plan (*Fig 9.03*) it is a square hall of 50 ft (15.2 m) side, roofed with gargantuan sloping stone slabs, simulating the thatched roof of the original. Instead of timber posts to hold up the roof, there is an inner double row of massive stone columns. The peripheral columns are reduced to mere pilasters, the load of the roof being taken up by a wall of cyclopean stone blocks piled one over the other without the use of binding material. Some of the open panels were filled in years later with stone grills of incongruous designs.

In the middle of the rear wall of the main hall a square portion was partitioned off to create a workable equivalent of an inner-sanctum, the *garbhagriha*. The peripheral timber bench with an inclined back (*asana*) that was a functional requirement of the 'santhagar'—was dispensed with in the main hall. In the attached portico, however, where the worshippers could sit and chat after their ritual was over, even the bench has been reproduced in stone, thereby serving the dual function of a resting place and a baluster. This element of folk architecture in fact was destined to become a major design element in the hands of the classic Hindu architect, particularly in the central and western regions of the country.

➤ *Facing page*
Fig 9.02 The AD 450 temple of Ladkhan as it stands today with its mandapa openings filled in with jaalis

*Fig 9.03 Ladkhana temple. Showing **1** Garbhagriha, **2** Mandapa and **3** Entrance portico*

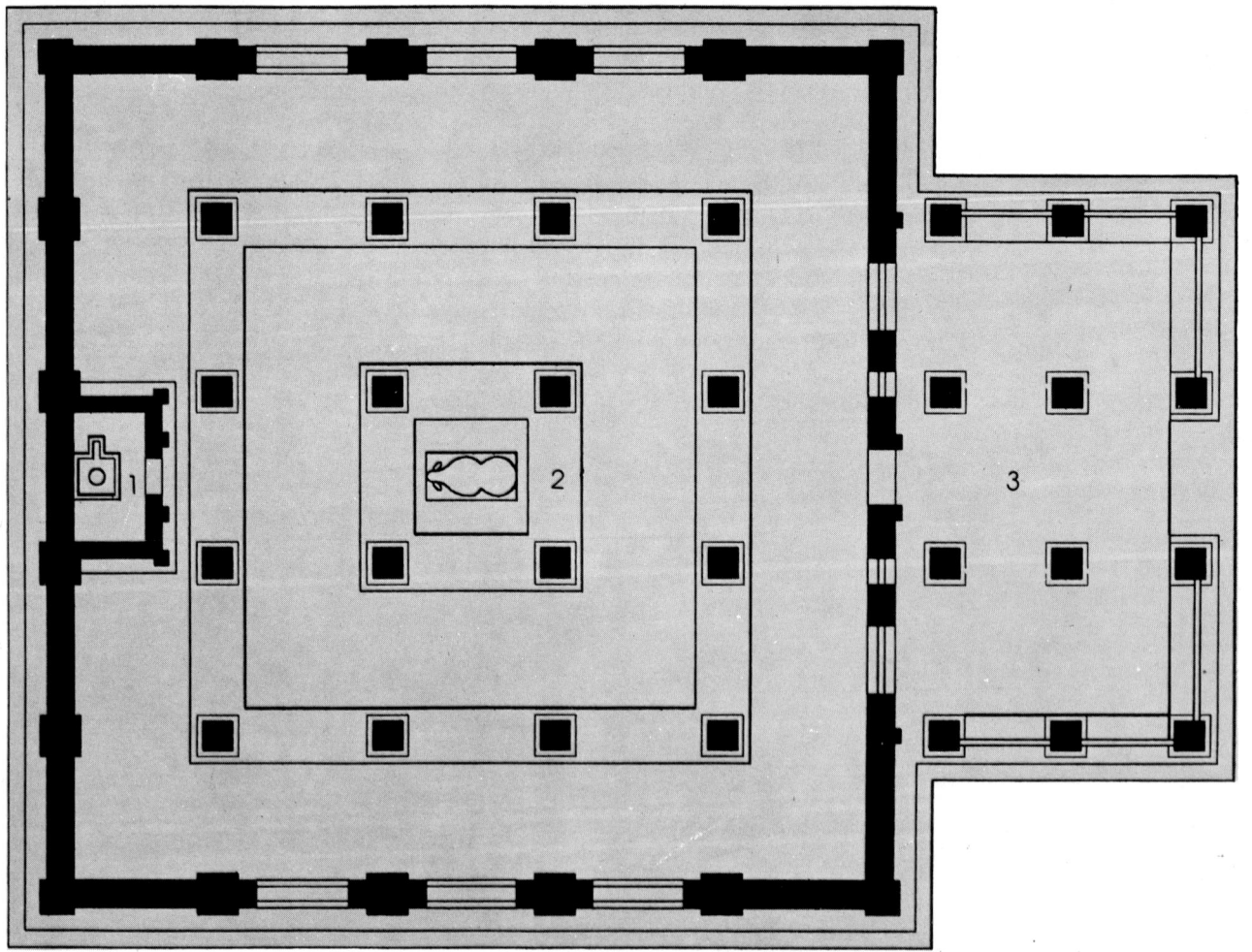

The Chalukyan builders copied the timber prototype of their choice as faithfully as their Buddhist predecessors. The projecting edges of the slabs of the roof are rounded off to imitate the original thatch overhang, and even the bamboo ribs holding the thatch down are reproduced in bulky semi-cylinders of stone (*Fig 9.04*). The result is a building of stark, strong and enduring appearance, relieved but a little by the iconography that displays the unmistakable plastic and sensual grace of the Gupta sculpture in Ajanta. To the Hindu mind already well versed in abstractions, this much too pedantic verisimilitude interpretation of a timber form, was far from the perfect shrine.

Fig 9.04 The Ladkhana temple in bulky ponderous stone, its forms inspired by the thatch and bamboo Santhagara

The Durga Temple, Aihole

Judging from the nearby Durga temple (*Fig 9.05*), the Chalukyan builders in their search for an appropriate monumental shrine tried to adapt the plan of the Buddhist *chaitya* hall to Hindu worship. In building this temple of Durga, they merely divided the inner colonnaded nave of the *chaitya* hall into an ante-chamber and sanctuary by putting a platform across near the apsidel end (*Fig 9.06*). The aisles around became an open *pradakshinapath* of sorts. The roof of cumbersome sloping stone slabs displays little advancement over the constructive techniques of the Lad Khan, though this temple was built years later. This production of the struggling Chalukyan builder was too obviously derived in all its aspects from the Buddhist *chaitya* hall. No wonder then that it failed to win Hindu approval as an appropriate sacrosanct edifice. It is apparent that the builders, inexperienced in using stone as a structural material, were unable to create the monuments that the Hindus desired to erect over their cavern of worship. They interpreted the Hindu concept of a temple as a 'house of God' rather too literally as just a house, and no more. What was apparently missing in these early shrines was an architectonic form that would at once adequately symbolize a new faith and convert the 'house' into a virtual monument.

Fig 9.06 The apsidal ended Buddhist chaitya hall adopted for Hindu worship by creating a garbhagriha where the stupa would have been

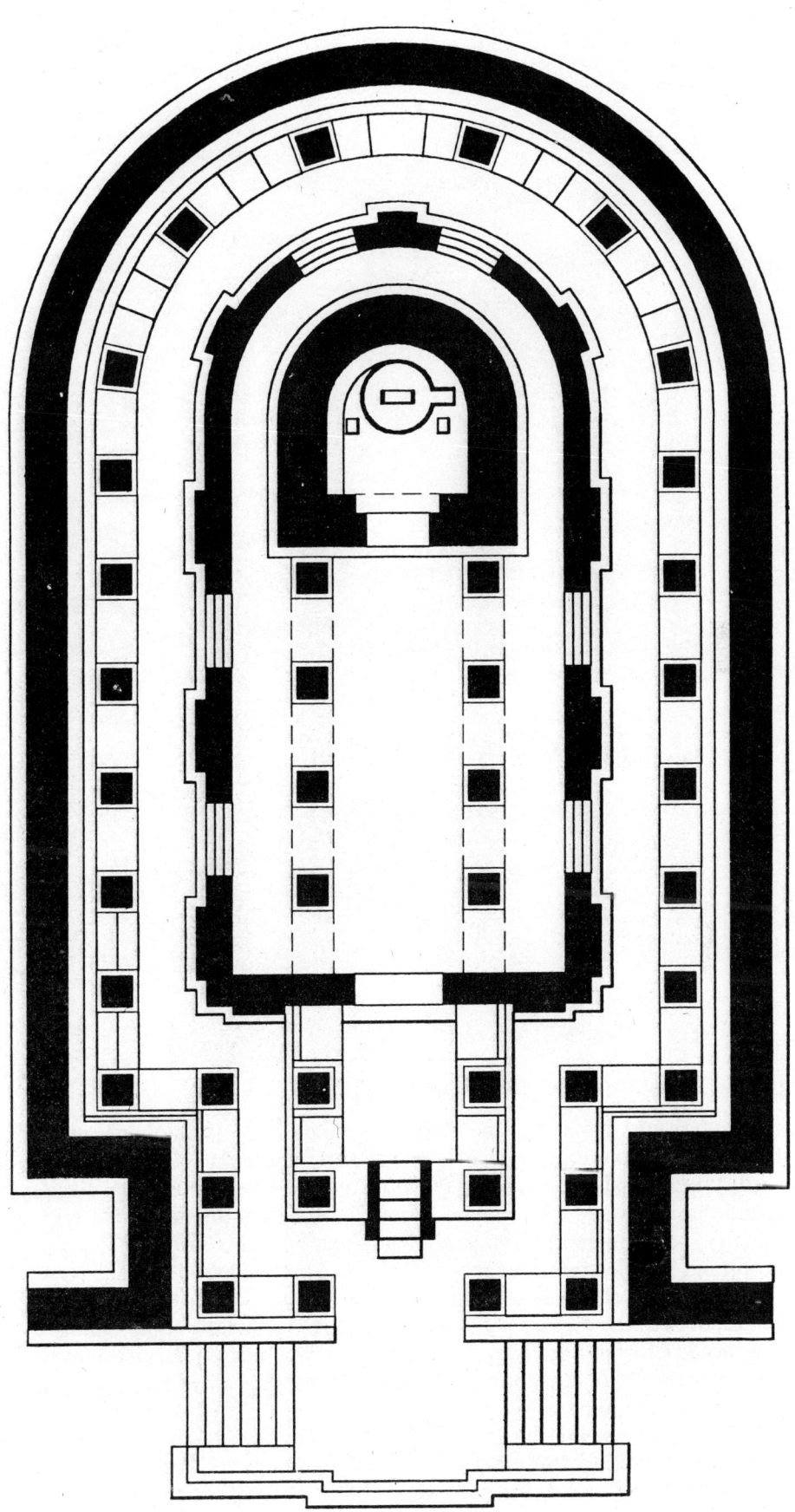

Fig 9.05 The AD 500 Durga temple at Aihole, planned like a Buddhist chaitya hall surrounded by a verandah with gargantaun pillars

Sophisticated Urbanity under the Guptas

One such symbol was destined to be developed miles away in the north, where under the Guptas a luxurious style of city life was emerging with most of the paraphernalia of towns and cities familiar to us today. The prosperous urban dweller had acquired the necessary wherewithal to live the life of a dilettante. He was a lover of poetry, painting and music, paid great attention to his sartorial elegance, collected objets d' art of ceramic, copper and iron, and patronized courtesans versed in the refined arts as well as the techniques of love expounded in the now famous Hindu treatise on sex, the *Kamasutra*. Towns and cities were developing rapidly to cater to the needs of such a society. Though society was not yet divided into the rigid Hindu caste system of later times, a horizontal stratification by economic status or occupation was beginning to tell.

The affluent section of the city was planned as a series of square or rectangular blocks allocated to the different classes or castes, with houses of varying heights "having balconies or belvederes made of wood" built around open courtyards. The shopping streets had houses for the owners or upper floors; a pattern still followed in the modern Indian bazaars of today. On the outskirts of the planned city "butchers, dancers, executioners, scavengers and so on have their abodes," made of wattled bamboo much like the shanty towns of today. The larger part of the city was built in timber and brick. The former was abundantly available from the forests around, and the sandy clay of the plains was ideal for the baking of the latter, using timber as fuel. These were also the most suitable materials for rapid and fairly durable construction.

Brick Temple of Bhitargaon

Over a period of time, bricklayers acquired a great degree of proficiency in their craft and were eventually called in by the Brahmin aristocracy to build temples for them. An early extant example of the bricklayer's skills, is a temple (*Fig 9.07*) at Bhitargaon (in modern Rampur district). It is built upon a substantial plinth and rises some 50 ft high. Structurally, it is an interesting building in its use of soundly constructed brick domes and vaults for its main cell and vestibule. The detailing of the surfaces with richly moulded surfaces (*Fig 9.08*) gradually ascending to a barrel-vaulted top is executed by craftsmen whose technical versatility is undeniable.

Fig 9.07 In plan this brick temple at Bhitargaon consists only of a 1 Garbhagriha and 2 Entrance portico

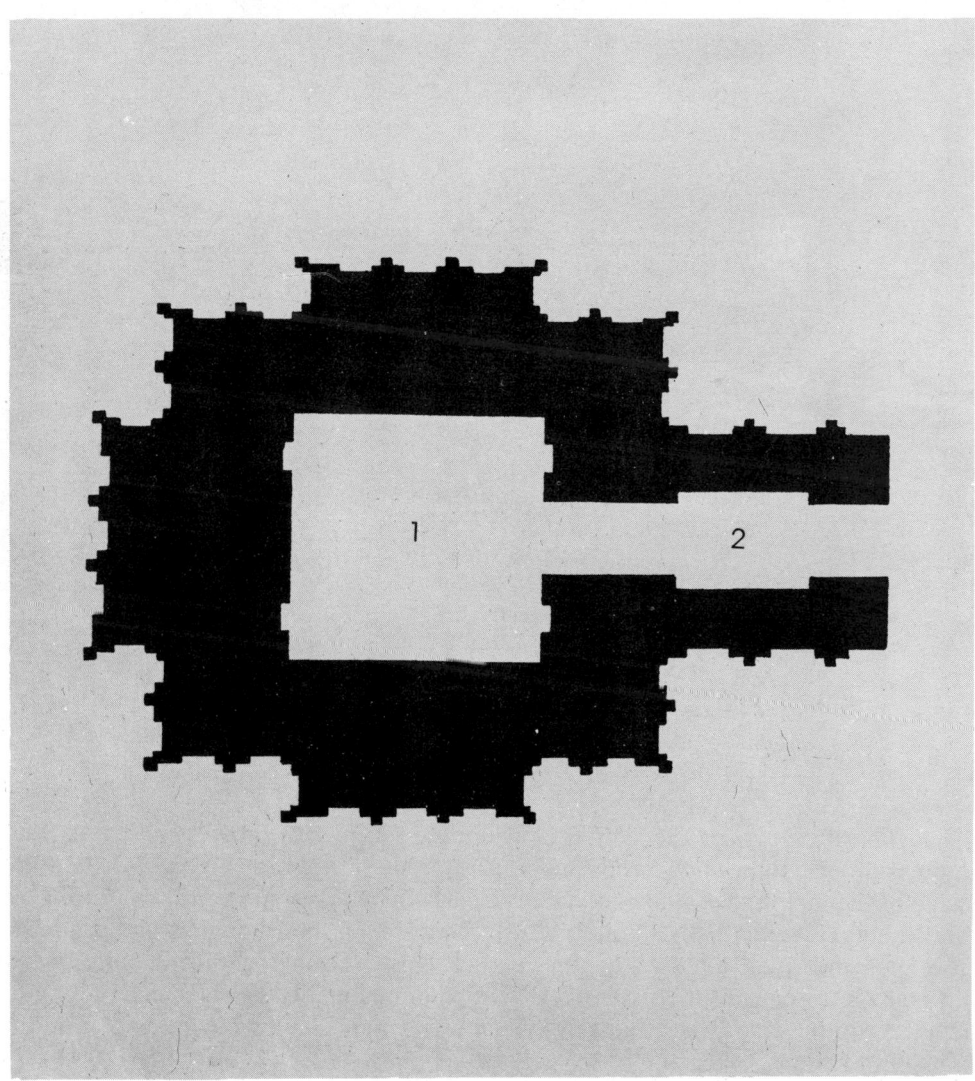

Fig 9.08 The brick layers treated the facade much like stone carvers by cutting and embossing sculptural patterns into the masonry

The silhouette of the Bhitargaon temple (*Fig 9.09*), however, leaves the impression of a rather ponderous and ungainly pile of masonry. What is more, the shape of the mound over the main cell is clearly inspired by the barrel vault roof of the Buddhist *chaitya* hall, a building form which even the Buddhists had dropped from their priorities, almost a century earlier. However ungraceful the building, it did manage to convey the essential idea of monumentality by its sheer bulk and height. Over the years, the Hindu craftsman was slowly creeping towards a solution to his problem.

*Fig 9.09 Bhitargaon temple of
AD 550 an example of the brick
mason's attempts at creating an
appropriate towering mass for
the Hindu temple*

Evolution of the Shikhara

It surely was some anonymous genius who, with a deep insight into the Aryan Hindu mind and an equally profound appreciation of its architectural potential, came up with the form of the parabolic profiled *shikhara*. The many variations of this pinnacle devised over the ensuing years, adorn all Hindu places of worship in the north that have any pretensions to being a temple, to date. Whether the inspiration for the form of the *shikhara* came from the profile of the traditional mythological abode of the gods, the snow-clad peaks of the Himalayas, or from the primitive megaliths or Toda churches of central India (*Fig 9.10a*), or from the shape of the traditional Indian wooden *ratha* (*Fig 9.10b*) or chariot, or from a rural Aryan form, since "all Indian art derives from the Aryan village," there is little doubt that the designer

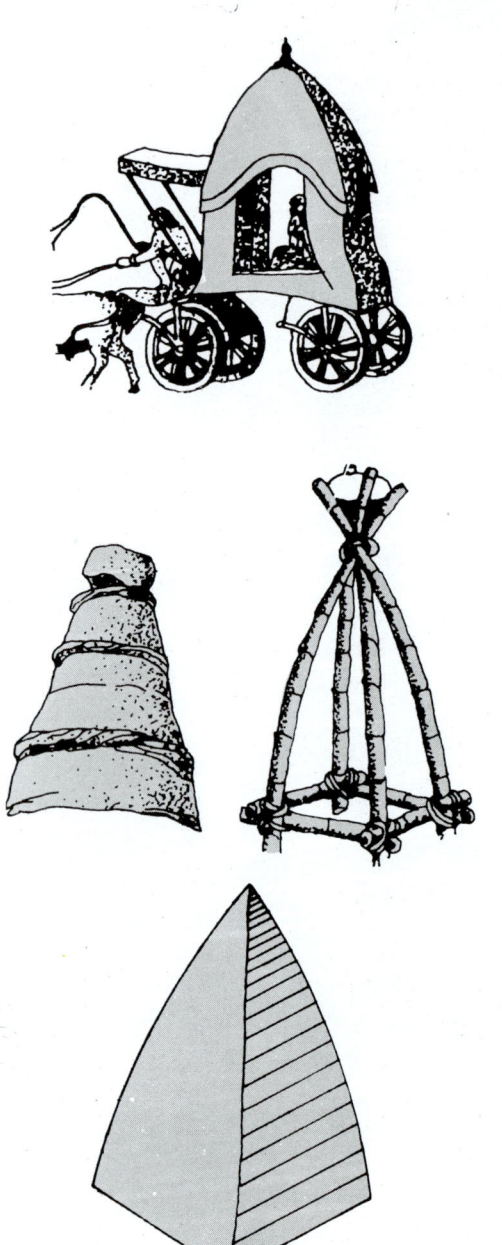

Figs 9.10 (a)–(d) The ratha, the toda church and bamboo roof that may have inspired the builders to create the form of the shikhara

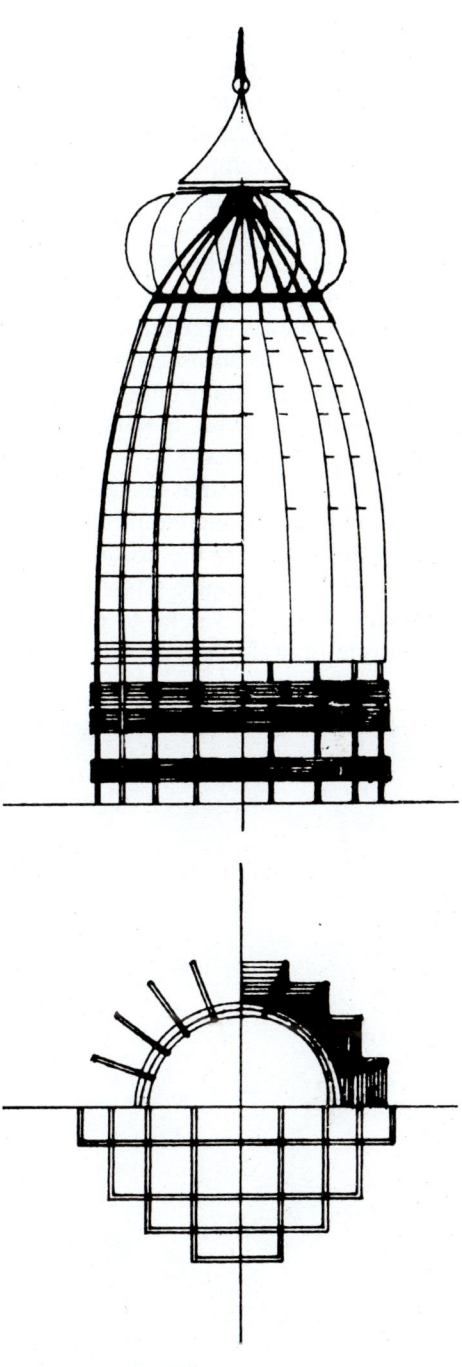

1

Fig 9.11 The famous shikhara, in its earliest glory. From all evidence it would seem that the earliest most perfect and symmetrically organised example of the shikhara, that was destined to crown the myriads of Hindu temples all over the country, was evolved in AD 500 at Deogarh in central India

came up with a form that is a puzzling and intriguing mixture of geometry and mythological symbolism. The simplest way to describe this form is as a four-sided pyramid with parabolic instead of straight edges (*Fig 9.10c*). Its curvilinear shape also strongly suggests its derivation from a bamboo construction, most likely that of four bamboos rising from the square base, held together at an appropriate height directly above the middle of the square (*Fig 9.10d*). It was decided to mount this parabolic form over the roof of the cube of the *garbhagriha* or the inner sanctum.

It is quite likely that initial experiments with this kind of shrine may well have been made with timber *shikharas* over dolmen shrines in stone. However, the earliest surviving classical form of it is, of course, a temple constructed in stone at Deogarh (*Fig 9.11*) in central India. A step by step evolution of its design is easy to follow.

The early Sanchi or Tigawa type dolmen shrine was the starting point. Complete with its attached portico, this was placed on a wide square platform inspired no doubt by the sacred sacrificial altars of the vedic Aryans. For reasons of symmetry, a false portico was attached to each of three otherwise blank walls of the cubic *garbhagriha*. What appear to be doorways are in fact porticos housing beautifully sculptured stone panels instead of timber shutters (*Fig 9.12*). The flat roof of the dolmen was now surmounted by an architecturalised form of convex profiled pyramid. This was built in discernible horizontal course of receding stone slabs approximating the cambered outline of the bamboo prototype. At the apex of this spire was mounted a large circular disc of stone into the rim of which was inscribed vertical edge as in an *amlaka* fruit—a fruit that plays a symbolic role in various Hindu rituals. This total composition, including the inevitable metal pinnacle or *kalasa* over the *amlaka*, is what came to be classified as the legendary *shikhara* (literally, mountain peak).

The designers had at last after years of endeavour evolved an acceptable new and distinguished architectonic symbol for the Hindu temple. The *shikhara*, by its more frequent usage in the north, was destined to become the distinctive feature of the North Indian temple. Classical writers in identifying the north with Aryan traditions conveniently refer to it as the most prominent feature of the Indo-Aryan temple style; a style that, as we shall see, was destined to give rise to the many

Fig 9.12 One of the three sculptured panels that adorn the false porticos of the Deogarh temple

CHAPTER 10

Harsha, the Last Great Buddhist

Fa-Hien, a Chinese Buddhist pilgrim of the Gupta period, traversed the entire length of India from west to east at the beginning of the fifth records with evident gratification that throughout his journey, extending over a decade, he was never once molested or attacked. Under the rule of the Guptas which lasted almost to the end of the fifth century, "India was perhaps the happiest and most civilized region of the world, for the effete Roman empire was nearing its destruction, and China was passing through a time of troubles."

Troubles across the borders, however, meant trouble for India as well. The central Asian nomads, this time the tribe of the Huns, were on the move again. The Gupta empire withstood their pressure long enough to divert the greatest of their fury towards Europe. Ultimately, in about AD 500, after defeating the Sassanians of Iran, Tormana the Hun established his capital in ancient Sagala, the former capital of King Menander.

Mihirgula the Hun

Menander's son, Mihirgula, a monster of iniquity to the Buddhists, turned upon them with great savagery, destroying many of their monastic establishments in the Gandhara region and slaughtering the inmates. Monasteries like the Takht-e-Bahi, which survived this holocaust, though in a ruinous state, have already been described.

Mihirgula's region of terror, though ferocious, was fortunately short-lived. A confederacy of Hindu kings led by Yashodharaman of Kanauj, finally forced him to seek shelter in the valley of Kashmir, where he died before he was able to renew his attacks in the plains below. And, once again, for over a hundred years India succumbed to the turmoils of a long period of unproductive feudalism. Finally, Prince Harsha of Thanesar (north of modern Delhi), through a series of fortuitous circumstances ascended the throne of Kanauj in AD 630. Over the next six years Harsha "went from east to west subduing all who were not obedient." After this incessant campaign, much like Asoka, he "reigned in peace for thirty years without striking a blow."

There is probably some truth in the story that Harsha, when a lad of sixteen, had entertained ideas of entering a Buddhist monastery. Circumstances catapulted him into accepting the responsibilities of a Hindu Kshatriya king, but his faith in the middle path shown by Buddha, was obviously unshaken. It was under his reign that Buddhism had its last days of glory in India. By this time, the religion had gained a large following not only in the north-western region of modern Afghanistan, but also eastwards in Burma, South Asia, and northwards across the Himalayas in China. India, and particularly the region of modern Bihar (evident from the many Buddhist *viharas* that dotted the countryside of the region) was to the Buddhist world at large, like Jerusalem to the Christians. Pilgrims and scholars were beginning to pour into India in search of solace or knowledge. The phenomena saw the blossoming of places associated with historic incidents in Buddha's life into centres of holy pilgrimage. Most popular of these are places like Kapilavastu, the place of the nativity, Budh Gaya, the scene of enlightenment, and Sarnath where the first sermon was delivered. Centres of learning such as that at Nalanda established years ago were now assuming worldwide importance, with students from far away lands seeking theological training in the land of the Buddha's birth.

Budh Gaya and Sarnath

The two great centres of pilgrimage were Budh Gaya and Sarnath. Sarnath (near modern Varanasi) at one time was a thriving and bustling monastic complex. A number of *stupas, stambhas, toranas* and many storeyed *viharas* accumulated on this holy site over hundreds of years, much like at Sanchi, unguided it would seem by an overall master plan of any sort (*Fig 10.01*). The Mahayana School having acquired prominence, the emphasis was shifting from *chaityas* to temples built after the fashion of Hindu shrines, since instead of worshipping the *stupa*, the image of Buddha itself was venerated, not unlike the craven image of a Hindu god.

Fig 10.01 The accretion of stupas, temples and monasteries at Sarnath, where Buddha hundreds of years earlier, had given his first sermon

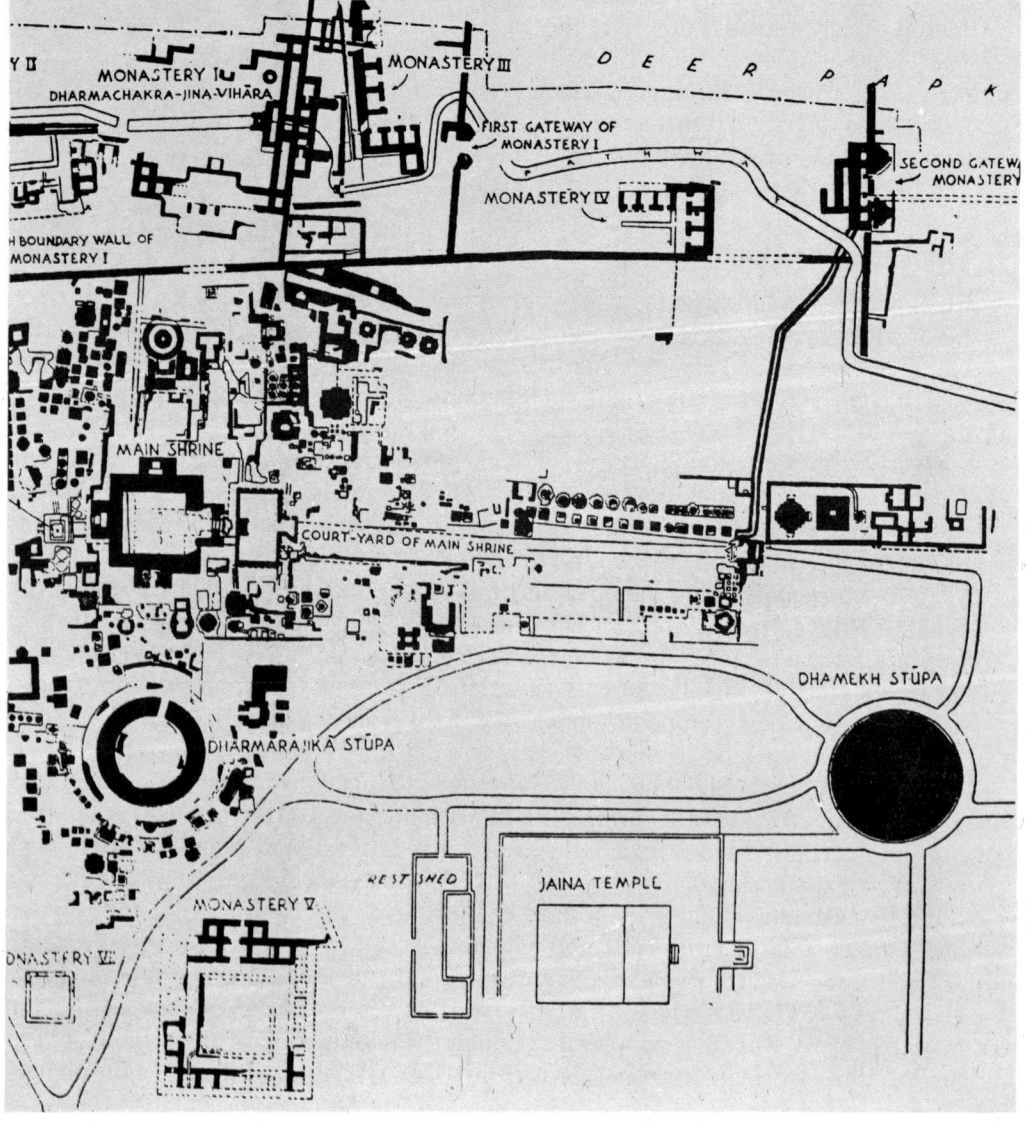

The tower of the temple at Sarnath, with its seven clearly marked receding storeys, rising to a height of 110 ft (33.4 m) derives inspiration both from the elongated stupas of Gandhara as well as the *shikharas* of Hindu shrines. At the apex of the pyramid was mounted a miniature stupa, and *harmika* pinnacle, to distinguish it from a Hindu temple. The nearby Dhamek stupa (*Fig 10.02*), now in a ruinous state, on the other hand, is more distinctly derived from the style of Gandhara. The hemispherical mound of brick was raised over a tall cylindrical base. The four cardinal points of the substructure are decorated with pilastered and arched niches which at one time were filled with large images of Buddha.

Fig 10.02 The form of the Dhamek stupa of AD 700 now in ruins was probably inspired by the Gandhara style raised as it is over a tall cylindrical base

The shrine of Budh Gaya (*Fig 10.03*) being a single large complex displays a more definite sense of design. The 180 ft (54.7 m) high temple spire (*Fig 10.04*) is the focus of the rather inconsequential modern structure around it. This temple was restored under the supervision of General Cunningham of the Archaeological Survey of India in 1881. Much of the original, which may have been centuries older, has been covered up. The restoration featured the building of a broad terrace at the base of the temple. At the corners of the platform, towers echoing the shape of the main spire were built up to conform to the *pancharatna* (five jewels) concept of planning. General Cunningham's work has been the subject of heated controversy among historians and archaeologists. In its original form, the temple, without the corner towers, is conjectured to have been cordoned off only by the traditional *vedika*, with may be a *torana* to mark the entrance. The shape of the original temple tower was probably not substantially different from that of Sarnath, described earlier.

Fig 10.03 The plan of the Bodh Gaya shrine is akin to that of a Hindu temple though it has all the symbolic trappings of a Buddhist shrine

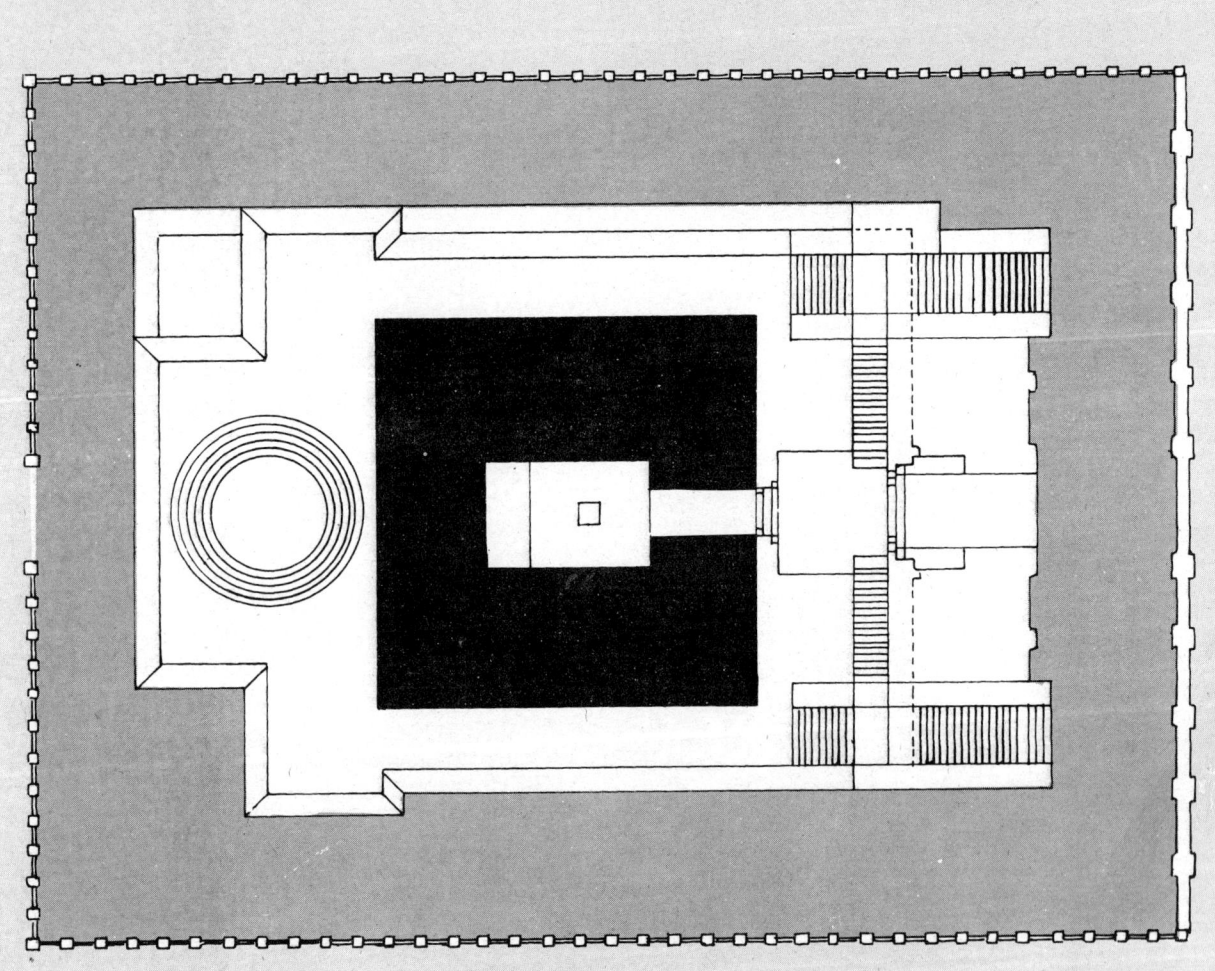

Fig 10.04 The spire of the Bodh Gaya temple, rebuilding of which in 1881, caused great controversy regarding the original design and General Cunningham's restoration

The Great University of Nalanda

For architectural magnificence, however, in its heyday the Buddhist university at Nalanda, near modern Patna, overshadowed both Sarnath and Budh Gaya. The most prominent feature at Nalanda was a massive stupa (*Fig 10.05*) of which only the most incomprehensible ruins of the successive accretions are extant. When complete, it was more than 110 ft (33.4 m) high and rested on a 30 ft (9.1 m) high, 100 ft (30.4 m) square platform. It was approached by a broad imposing flight of steps rising up from a central avenue. The walls of the base of the stupa were divided with cornices into horizontal friezes, further partitioned vertically into pilastered niches containing either the sun window emblem or images of Buddha. At the corners of the main platform were buttresses in the form of turrets crowned by mini-stupas. This *pancharatna* concept of planning consisting of a central shrine and four subsidiary ones at the corners (*Fig 10.06*) became the essence of design of the Buddhist style of architecture that flourished subsequently in south-east Asia. The famous temple at Angkor Vat in Cambodia is the most glorious of these.

➢ *Facing page*
Fig 10.05 The brick remains of the splendid stupa that was the monument magnificent of the once great Nalanda establishment, of about AD 500

Fig 10.06 The Angkor Vat in Cambodia, that was the culmination of the panchratna concept of Buddhist planning

The Planning of the Campus

Rows of such magnificent stupas placed side by side formed the central monumental axis of the Nalanda complex (*Fig 10.07*). Parallel to, and on both sides of this axis were what may be called 'avenues of the monasteries', since on either side of these were establishments housing students aspiring to become monks. These monasteries or *viharas* (a heap of ruins now, built as they were in brick and timber) were planned in the usual manner around a central open court. They functioned both as units of residence as well as learning. The lower storeys contained the refectory, areas of instructions and communal worship, while the upper floor built in timber was a quadrangle of cells for resident students. Each subsequent higher storey was stepped back from the previous one to create open-air terraces for the cells. Also, it is said, the higher storeys with smaller number of cells housed the more persevering students who had graduated to an advanced stage of learning.

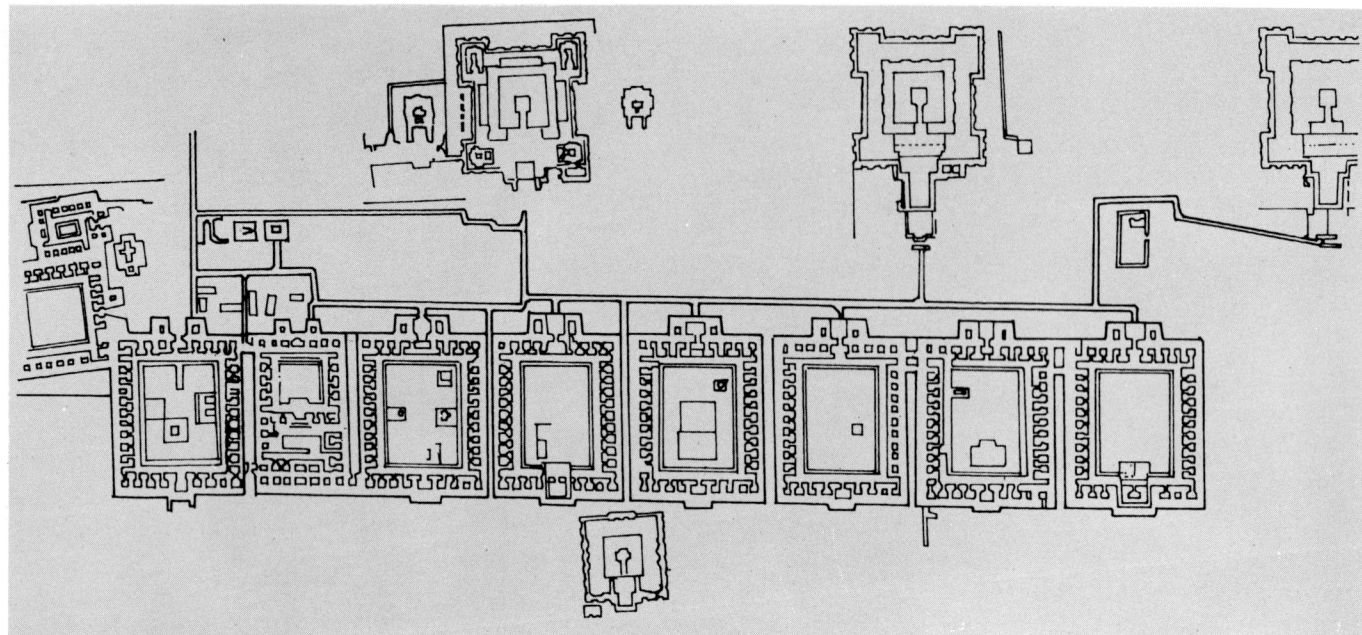

A number of such monasteries were built close to each other like so many colleges in a university campus. Complemented by the white and gilded domes of the stupa rising into the sky, they were seen by the famous Chinese pilgrim Hiuan Tsang in the seventh century before they were razed to the ground by the Muslim invaders of the eleventh century. "The richly adorned towers, and the fairy-like turrets," he says, "like pointed hilltops are congregated together. The observatories seem to be lost in vapours of the morning, and the upper rooms tower above the clouds. The outside courts in which are the priest chambers are in four stages and have dragon projections and coloured eaves, the pearl red pillars carved and ornamented, the richly adorned balustrades and roofs covered with tiles that reflect the light in a thousand ways, these things add to the beauty of the scene." This, however, as we shall see, was destined to be the last 'beautiful scene' of Buddhist architecture in India. While Buddhism took root in the various eastern and south-eastern countries to blossom into a creed that is still a living religion, in India the psychological onslaught of the Brahmins ultimately prevailed. The history of Indian architecture for over 500 years hereafter is virtually the story of the blossoming of the Hindu temple into great cathedrals.

Fig 10.07 The orderly layout of "colleges", hostels and dormitories along the classic avenues of learning that were the core of the campus of the Nalanda university which in its heyday was "richly adorned by towers and fairy-like turrets"

CHAPTER 11

The Chalukya–Pallava Power Balance

Fig 11.01 (a) An overall view of the ancient town of Badami

Aihole remained the capital of the Chalukya kings, till the first Pulakesin's son, Mangalisa, shifted it to nearby Vitapi (modern Badami) in AD 578. Aihole, as we have seen, was the site of the early experiments in Hindu temple building which had resulted in the rather lacklustre shrines of Lad Khan and Durga. Badami, the new capital, was picturesquely located on a lake, surrounded by steeply rising cliffs (*Fig 11.01a*) like those of the Western Ghats and Ajanta.

Rock-Cutting at Badami

As was to be expected, rock-cutters were soon at work carving out halls from the scarp of hill overlooking the south-east of the town. Like their predecessors at Aihole, their notions of the appropriate form for a Hindu shrine were quite vague. The typical Hindu and Jain cave sanctum of Badami (*Fig 11.01b*), therefore, turned out to be nothing more than an outer pillared-veranda, through which one entered a colonnaded hall to worship the Shiva-linga in a cell of sorts, right at the back. The builders had done no more than make 'additions and alterations' to the plan of an Ajanta *vihara* to make it into a workable Hindu shrine. While the rock-cutters patiently carved out a row of such, rather insipid, cave temples, the never-say-die stone masons of Aihole turned their attentions to adding lustre to the architectural qualities of the rather nondescript existing edifices—the well known Lad Khan and Durga temples at Aihole.

Additions to Durga and Lad Khan

In the Durga temple, the eminently successful *shikhara*, the form of which had already been developed in the north, was rather incongruously planted over the apsidal end of the flat roof (*Fig 11.02*). The Lad Khan, on the other hand, was adorned with a cubic volume built over the central flat portion of the roof over the main hall (*Fig 11.03*). In both cases, no doubt the height of the original was enhanced. This alone, however, could not be expected to produce an architectural composition of distinctive quality. Nevertheless, the latter effort proved to be not entirely in vain. The design of an old temple, now virtually in ruins, on a ledge of rock overlooking Badami, is quite likely an inspired though radically modified version of the profiles of the Lad Khan.

➢ *Facing page*
Fig 11.01 (b) A facade of the rock-cut Hindu cave at Badami

Fig 11.02 The shikhara having been accepted as the appropriate symbol of Hindu faith was planted over existing temples such as the Durga at Aihole

Fig 11.03 *Instead of the shikhara, a cubic volume was planted over the Lad Khan in an attempt to give height and architectural eminence to a rather ordinary shrine*

The Evolution of the *Vimana*

The designer of this unnamed temple (*Fig 11.04a*) seems to have reorganized the rather disparate elements of the Lad Khan into an integrated whole. The *garbhagriha* which had been an incongruously partitioned space in the rear of the Lad Khan, is appropriately relocated at the core of the composition. The aisles so formed around the cella became an enclosed *pradakshinapath*, roofed though, like in its prototype, with massive sloping slabs of stone. The ground plan (*Fig 11.04b*) is completed by the addition of a small *mandapa* or ante-room preceding the *garbhagriha*, and an entrance portico similar to that of the Lad Khan.

It is, however, the design of the dolmen addition over the *garbhagriha* (*Fig 11.04c*) and the conscious shedding of obvious folk art trappings of the Lad Khan temple that mark a decisive step in the evolution of classical temple architecture in the south. The tower now takes the form of a stepped pyramid, every step defined by horizontal moulding of varying thickness. Each one, though, is rounded off, as if to consciously remind the viewer of its origins from the gentle overhang of the thatch roofs over bamboo huts. The apex is crowned by a domical form, derived as ever from a rural one. In this case, the architects preferred the shallow, almost semicircular contours of a bamboo canopy, built up over a square base (*Fig 11.04d*). With minimal modifications this curvilinear form was hewn out of massive blocks of masonry and planted squat over the tower.

Fig 11.04a

Figs 11.04 (a)–(d) The plan, sectional restored view and other details of the ruined temple at Badami, that forms an essential link in tracing the design development of the South Indian vimana

Fig 11.04b

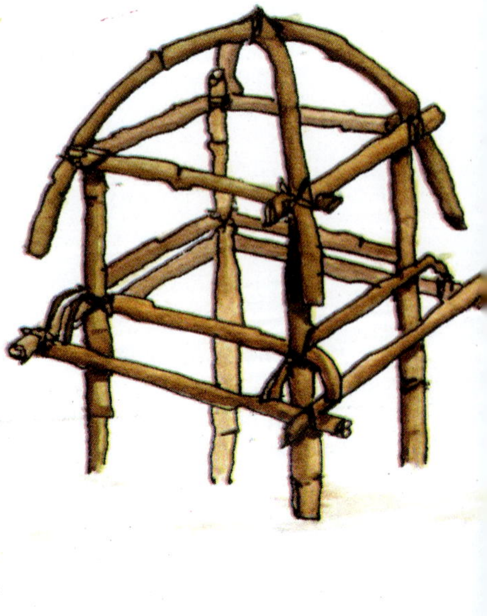

The combination of a stepped pyramid and dome with a cubic or prismatic base, even in its primitive form, was a novel and pleasing ensemble. It was to become the hallmark of south Indian temple architecture, symbolizing the Shaivite aspect of the Hindu trinity which enjoyed great popularity in the South just as the Vaishnavite type enjoyed in the North.

The pyramidal tower (popularly known as the *vimana* in south Indian architectural terminology) crowning the temple of Badami, though, was far from perfect. The inexperienced hands of the stone mason, working with large unwieldy blocks of stone had given birth to a new concept, but refinements of detail and the fluidity of the profile of the *vimana* was to be achieved by another group of craftsmen elsewhere in India.

Fig 11.04d

Fig 11.04c

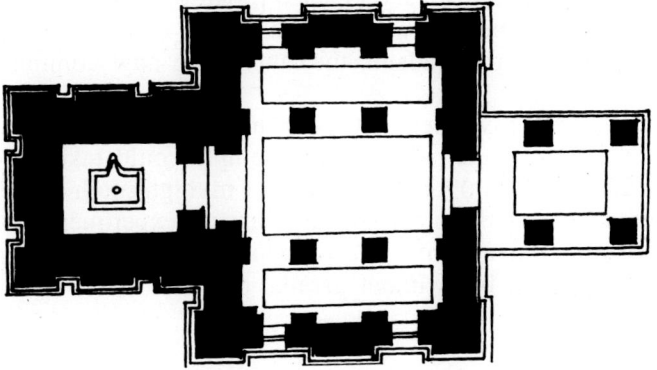

The Migrant Craftsmen

Meanwhile, Chalukyan control over Vengi, a large consolidated tract of cultivatable flat land lying between the Krishna and Godavari rivers, was challenged by the Pallavas of the upper southern coast. The powers of the Chalukyas and Pallavas, however, were evenly balanced. This resulted in frequent tests of strength between these two major powers of the south. The Pandya, Kadamba and Ganga dynasties of the further south were content to play second fiddle. They supported either side to suit their own political expediency, without losing control over their own smaller kingdoms.

At the conclusion of such battles, skilled craftsmen either transferred their allegiance voluntarily or were often forcibly carried away from the kingdom of the vanquished to built temples for the victorious king. During this period there was, therefore, a parallel and interrelated growth of temple architecture in the cities of the Chalukyas and Pallavas, since the masons and sculptors who were shunted back and forth belonged to the same school of craftsmanship.

The Pallavas and Mahabalipuram

Mahabalipuram was the coastal capital of the early Pallavas. Situated at the mouth of the Palar river, it stood on a 100 ft (33.3 m) high, and a half mile (800 m) long hill of granite gneiss rising out of the surrounding sands. The citadel on top of the mound was secured by impressive fortifications. The administration offices and residences for merchants and members of the ruling class, located around the central place, were built over stone basements, some of which have survived to date. The superstructures, presumably of timber and brick, have vanished. The remnants of water channels and conduits show that the town was served by an elaborate and well designed system of water supply and drainage.

Mahendra Varman II, one of the Pallava rulers of the early seventh century, and a contemporary of Harshvardhana of Thanesar, was the major figure responsible for the political and cultural growth of the Tamil country. Though born a Jaina, he was converted to Shaivism quite early in his life. He displayed an unusual interest in the arts, being himself a poet and even the author of a play entitled "The Delights of the Drunkard" (*Mattavilass Prahasana*). It was under his patronage that a novel and significant experiment in the art of temple building in the south was carried out.

Guilds of Masons and Cutters

It is quite likely that by now two distinct guilds of building craftsmen had evolved; that of stone masons and stone carvers, each proud of its own tradition and even eager to challenge the other. Aware of the aesthetic shortcomings of the efforts of the former at Aihole and Badami, the latter were now commissioned by the Pallavas.

In an attempt to prove their virtuosity they created in one spot a multitude of forms and shapes that the 'ideal' Hindu temple could take. A small granite outcrop, about 250 ft (75.2 m) long and 30 ft (9.1 m) high south of the fortifications of Mahabalipuram became the site of the most unusual experiment in the techniques of building, resulting in a form of architecture that flourished into a mature consummate art in India, but without parallel outside.

Free Standing Monoliths

Instead of carving caves out of the living rock as his forebears had done for centuries in India, the sculptor architect, with surprising aplomb proceeded to chisel down the granite outcrop into free-standing monolithic models of structural buildings, much like the romantic gardener pares thick hedges into shapes of animals and birds. The stone carvers' perseverance resulted in the famous seven pagodas or *rathas* (literally, chariots) of Mahabalipuram (*Fig 11.05*). Due to the disposition of the outcrop—a narrow and long whale-backed hump of none too great dimension— the seven virtual architectural models are of moderate size and aligned along a single axis. Though for some obscure historic reasons, the work of finishing and polishing 'the models' was abandoned, the artists managed to present to the king an array of distinctly comprehensible temple forms. Each of these eight was subsequently named after the heroes of the great Hindu epic, the *Mahabharata*.

Fig 11.05 The Pallava builders added an altogether new dimension to the art of rock cutting in India, when in 650 AD they chose to cut down an outcrop of living rock into rathas of various shapes and designs—the famous seven rathas of Mahabalipuram

The Rathas of Mahabalipuram

Draupadi (*Fig 11.06a*), the simplest of the so-called *rathas*, is an unpretentious version of the rural portable wooden shrine with a square base and curvilinear thatch roof. Three others are varying interpretations of the Buddhist *chaitya* hall. The most elementary of these is rectangular in plan with a peripheral colonnade covered with a barrel-vaulted roof; yet another, the Sahadeva (*Fig 11.06b*) mounts a three-tiered roof over an apsidal-ended plan, culminating in the barrel-vaulted *chaitya* form at the apex. In the Ganesh Ratha (*Fig 11.06c*), a rectangular entrance portico is attached to the already complex ensemble of forms of the Sahadeva Ratha.

Figs 11.06 (a)–(d) The Draupadi, Sahadeva, Ganesh and Bhim rathas at Mahabalipuram

Fig 11.06a

Fig 11.06b

Fig 11.06c

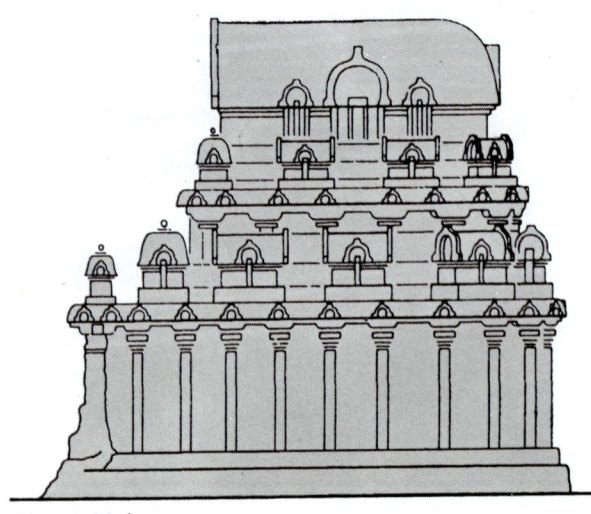

Fig 11.06d

Though the forms of the Bhima, Sahadeva and Ganesh *ratha* were also absorbed into the myriads of forms of the later temple architecture of the south, the form of the Dharmaraj proved to be the 'king' among all (*Fig 11.06d*). Built over a square plan, it was found to do the greatest aesthetic justice to all the formal, cosmological and ritual requirements of a temple, as explained in detail in a later chapter on the theories of Hindu architecture. Each receding level (*bhumi*) of the pyramidal roof of the Dharamaraj is in itself a miniature model of the Buddhist *vihara* (*Fig 11.07a, b*). What were cells for the residence of the monk in the original, are here frozen into either miniature ornamental domes or barrel-vaulted turrets. The place of the central open court of the living *vihara* is taken up by a solid core of rectangular masonry diminishing in size as it approaches the apex. The finial of this finely proportioned and meticulously carved pyramid is an elegantly contoured ribbed octagonal dome, poised over a squat cylindrical shaft. The whole, in fact, is a perfectly sculptured crystallization of the rather cumbersome structural efforts of the builders of Badami.

The Chalukyan builder and the Pallava sculptor had thus between themselves evolved an appropriately aesthetic spire (*vimana*) for the *garbhagriha* of the South Indian temple.

Fig 11.07 (a) Dharmaraja the supreme ratha that was ultimately adopted as the most appropriate form of the vimana to crown the South Indian Hindu temple

Fig 11.07 (b) Dharmaraja,
another view

CHAPTER 12

The Triumphant Pallavas and Rashtrakutas

Flushed with the triumph of his artists, Mahendra Varman's successor Rajasimha, proceeded to dot the eastern coast line and his new seventh century capital of Kanchipuram with "structive" temples, that were variations on the Dharmaraja theme.

Shore Temple, Mahabalipuram

The most exquisite and well preserved of these is the "shore temple" at Mahabalipuram (*Fig 12.01*). Delicately poised on the rocky shoreline, half in the sea and half on land, the main sanctuary of the temple looks out towards the sea while a subsidiary one faces the shore. Such an arrangement was necessitated by its peculiar location since the entrance to the former became flooded and unapproachable at high tide. The spires over either of the cellas are the earliest and probably the most serene and beautiful examples of soundly structured masonry versions of the sculptural Dharamaraja Ratha. Though the geometric scheme of these is the same as that of the Dharmaraja, their somewhat elongated profile is like a graceful feminine complement to the strong and robust masculinity of their Dharamaraja prototype.

Both the sanctuaries and subsidiary *mandapas* of the temple stood within a walled precinct (*Fig 12.02*). The courtyard of the temple could be partially flooded by a complex system of channels, reservoirs and drains that carried the sea water into basins around the cult room. The Pallavas, a sea-faring people, conducted rituals of worship of water in this temple. To the Pallava mariner, the edifice was a virtual lighthouse. He was guided to his haven of safety by observing a lamp lit on a pillar in the courtyard that could be seen from the ocean through an aperture in the wall. The quality of the construction of the shore temple marks the graduation of the South Indian stone mason from an initiate into an accomplished craftsman.

Fig 12.02 Plan of the shore temple at Mahabalipuram showing location of two shrines back to back

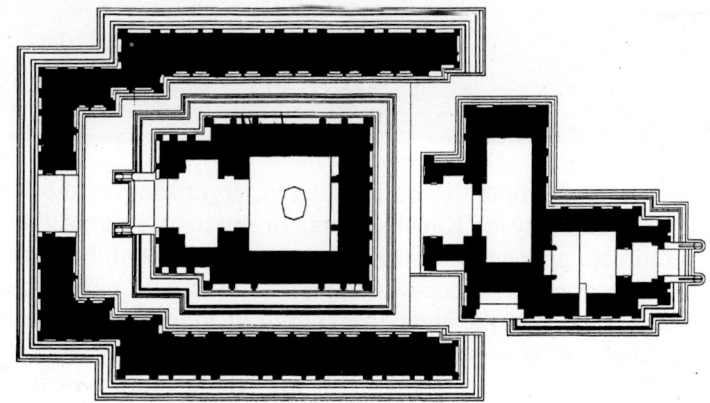

Rajasimha now attempted to build an even larger temple, the Kailash Nath, within his new capital city of Kanchipuram. In order to enlarge the base of the temple, the architects appended mini *rathas* at the corners and centre of the four walls of the cella; a none too successful experiment that was never again repeated. The contours and details of the stepped pyramid of the tower, however, conform to that of the Dharmaraja *vimana*, which obviously had won acceptance as the most prominent architectonic appendage of the Hindu temple in the South.

Fig 12.01 The shore temple at Mahabalipuram on the Eastern coast of India built half into the sea and half on land

Chalukyas Evolve an Integrated Plan

The Pallava craftsmen, by concentrating their sculptural skills on the shape of the *vimana* alone, had undoubtedly evolved an elegant principal form for it. To them other functional elements of the temple like the *mandapa* (assembly hall) and entrance portico were mere ancillaries. They had not yet been able to integrate these into a balanced architectural whole. The Chalukyan artist, developing on the experience gained in organizing disparate elements of the Lad Khan into the architectural entity of the temple of Badami displayed greater concern in evolving a logical overall architectural plan that would gather together the sanctum, *mandapa* and the entrance portico into a comprehensive planning entity. It is only natural then that a ground plan, reconciling the various components of the Hindu temple into an equable aggregate ultimately emerged in Chalukyan country. Under Vikramaditya I, the Chalukyans had established yet another capital city, that of Pattadakkal, situated half way between the older cities of Badami and Aihole. The seventh century Papanath temple at Pattadakkal (*Fig 12.03*) is the last example of a southern shrine adorned with a *shikhara* instead of the traditionally accepted vimana form. The craftsmen realized that the rather slender form of the *shikhara* could not substantially counterpoise the sprawling lower mass of the horizontality of their plan (*Fig 12.04*). Moreover, the Dharamaraja Rath form of the Pallava sculptors of Mahabalipuram had already won the enthusiastic approval of the priests and kings as the most appropriate temple spire profile.

Fig 12.03 The seventh century temple of Papanath at Pattadakal is the last of the temples in the Southern tradition to be crowned with a shikhara

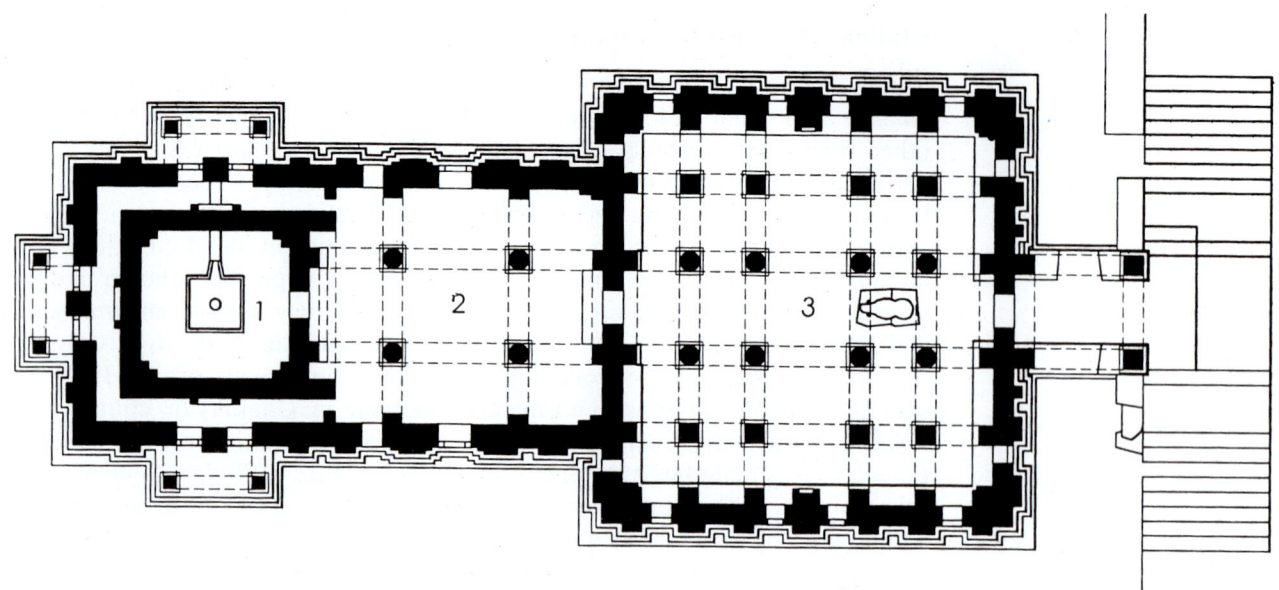

Fig 12.04 *Plan of the Papanath temple at Pattadakal.*
1 Garbhagriha, 2 Pradakshina, 3 Mandapa

The Virupaksha at Pattadakkal

In the temple of Virupaksha (*Fig 12.05*) built at Pattadakkal, in the immediate neighbourhood of Papanath, the Chalukyan builder decided to surmount the *garbhagriha* with the more acceptable form of the southern *vimana*. In striking contrast to the Papanath, it at once presents an altogether substantial silhouette. To achieve equable balance between the upper and the lower parts of the temple, a richly sculptured barrel-shaped form was projected out from the middle of the *vimana* to interlock as it were the tower and the *mandapas*. It effectively restrains too, the perceptible horizontal thrust of the mass below. The *mandapa* for the first time is not treated merely as just another hall attached to the cella; rather, a more convincing spatial link-up is established between the two by enveloping the dark square cella of the *garbhagriha* within the definable rectangular structure of the outer walls of the *mandapa* (*Fig 12.06*). The corridor-like space along the back and the sides of the inner sanctum thus becomes a virtual *pradakshina* or circumambulatory path for the worshipper. He offers his momentary obeisance to the gods from a narrow ante chamber like space between the opening to the cella and the clearly defined space of the *mandapa*. The *mandapa* in turn could be approached through any of the three porticos in the middle of its outer walls. As is apparent from the sum total, the superstructure of the temple is built up from a neatly synchronized ground plan.

Fig 12.05 Plan of the famous Virupaksha temple at Pattadakal. 1 Garbhagriha, 2 Pradakshina, 3 Mandapa

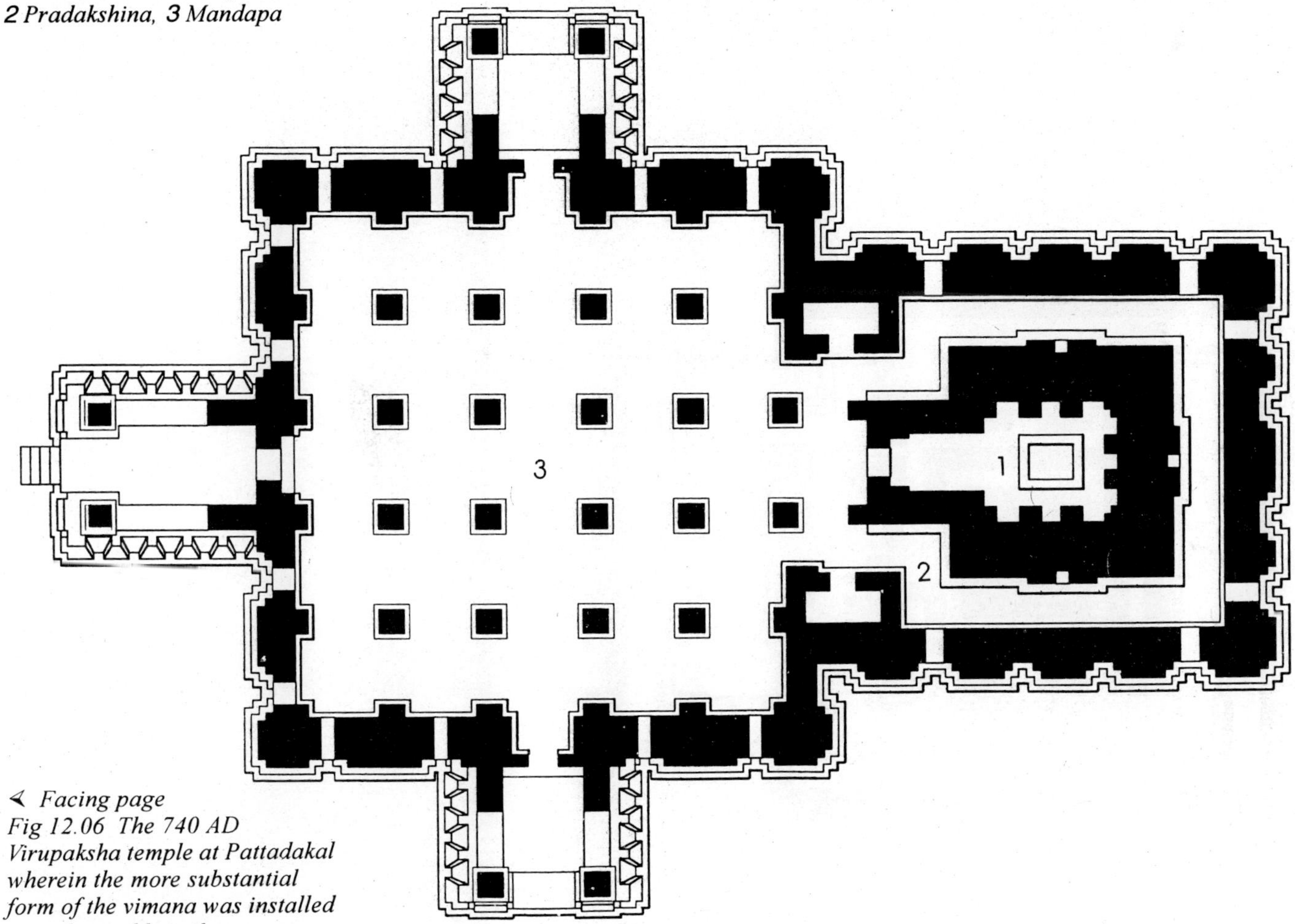

◁ *Facing page*
Fig 12.06 The 740 AD Virupaksha temple at Pattadakal wherein the more substantial form of the vimana was installed over the garbhagriha

A Touch of Pallava Elegance

It was the Pallava craftsmen, once again, who added the necessary touch of elegance and perfection to the robust matter-of-fact architecture of the Chalukyas. The germs of the idea embodied in the Virupaksha temple are carried to their fruition in the Vaikuntha Perumal temple (*Fig 12.07*) in the ancient city of Kanchipuram. Here, the entrance portico of the *mandapa* is dispensed with and greater body added to the mass of the cella by a pair instead of just one *pradakshinapath*. The diminishing form of the spire of the Vaikuntha Perumal is the only example of an Indian temple that in fact makes functional use of the substantial mass embodied in the profile of the *vimana*. The *vimana* interior is actually designed as three distinct usable storeys, each with its own central shrine (*Fig 12.08*). The two lower spires are surrounded by *pradakshinapaths* as well. The temple is virtually hugged by a colonnaded veranda a mere 8 ft (2.4 m) away and consistently parallel to the mass of the outer walls of the main temple. The inner surface of the unpunctured walls of the surrounding veranda is adorned with a large as life freely composed sculptured frieze glorifying the origins and history of the Pallava kings (*Fig 12.09*). The tightly knit plan, the intimate spaces, the multiple shrines, and the self-glorifying Pallava sculpture suggest that the Vaikuntha Perumal was not a public shrine; rather, it was the private family chapel of royalty, situated within or adjacent to the palace precinct.

Fig 12.07 Part plan of the Vaikuntha Perumal temple at Kanchipuram. 1 Garbhagirha, 2 Mandapa and 3 Pradakshina

➤ *Facing page*
Fig 12.08 The AD 720 Vaikuntha Perumal at Kanchipuram with receding stories and pradakshina path's embodied in the mass of the its vimana

Fig 12.09 The sculptured walls of the surrounding verandas of the Vaikuntha Perumal glorifying the temporal powers of the day

Fig 12.07

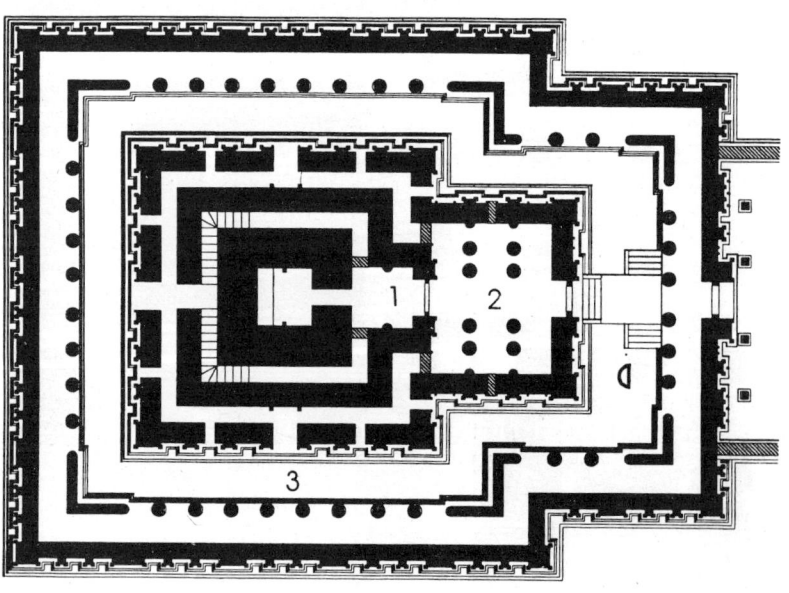

Struggles and Emergence of the Rashtrakutas

The halcyon days of Pallava royalty, however, were to end soon. The Chalukya king Vikramaditya was smarting under the blow dealt to his father by the Pallavas and soon gained sufficient strength to attach the Pallava capital of Kanchipuram in AD 740. Though he succeeded in decimating the Pallavas, he is said to have spared the cities. This proved to be the last great test of strength between the Pallavas and the Chalukyas. Both the powers had exhausted their wealth and energies in incessant mutual wars. Rulers of other dynasties were quick to take advantage of the ensuing power vacuum in the southern peninsula.

The Cholas of the deeper south began gnawing at the remaining regions of Pallava power. The Rashtrakutas, the heirs of the early Vakatakas, had until now focused their attention on the northern city of Kanauj. Now they found a rapidly weakening power on their southern flank. Doing a right about turn to suit the new situation, the Rashtrakutas quickly overcame the now feeble Chalukyan forces.

The heart of the new Rashtrakuta kingdom was the Western Ghats, a countryside in which the rock-cut form of architecture had flourished for over 700 years, since the time the Buddhist monks had carved out the caves of Ajanta and Karli. It was not surprising, therefore, that the art be rejuvenated under the growing power of the Rashtrakutas. The revival got underway with the excavations of cave temples on the ancient island of Ghorapuri (modern Elephanta), just off the shores of modern Bombay.

The Caves of Elephanta

In the closely clustered group of low hills, it was possible to carve out caves that were more like canopies of living rock hanging over generously lit spaces rather than mere tunnels of twilight run into the faces of cliffs as earlier caves had been. The *mandapa* which was approachable from three sides acquired a cruciform rather than rectangular plan of the earlier efforts at Badami, Ajanta and Mahabalipuram (*Fig 12.10*). The *mandapa* is spaced out by columns to form aisles of varying widths. The planners were probably again imitating some structural timber model faithfully since such variations in the size of aisles were redundant in the fashioning of caves, freed as they were of the restraints of structural spans. The Indian carver, avidly comprehending the sculptural potentialities of the dimly lit rear wall of the hall, was not in any case to be thwarted by the niceties of symmetrical planning. The enclosed shrine of Shiva, instead of being located in this more logical position at the rear of the hall, was placed instead in the middle of the entrance. This seemingly irrational choice is amply justified when we look at the rear wall. It is brought to life by the famous titanic sculpture of the Trimoorti "looming out of the unknown depths of the spaces and eternity" (*Fig 12.11*). In fact, it is obvious from the ponderous character of the massive columns, neither aligned to plumb nor accurately laid out to the right angle, that the accent in most Hindu cave temples was on creating surfaces for gigantic sculpture, rather than the spatial magic of Buddhist halls.

Fig 12.10 Plan of the great cave at Elephanta, near Bombay (AD 780)

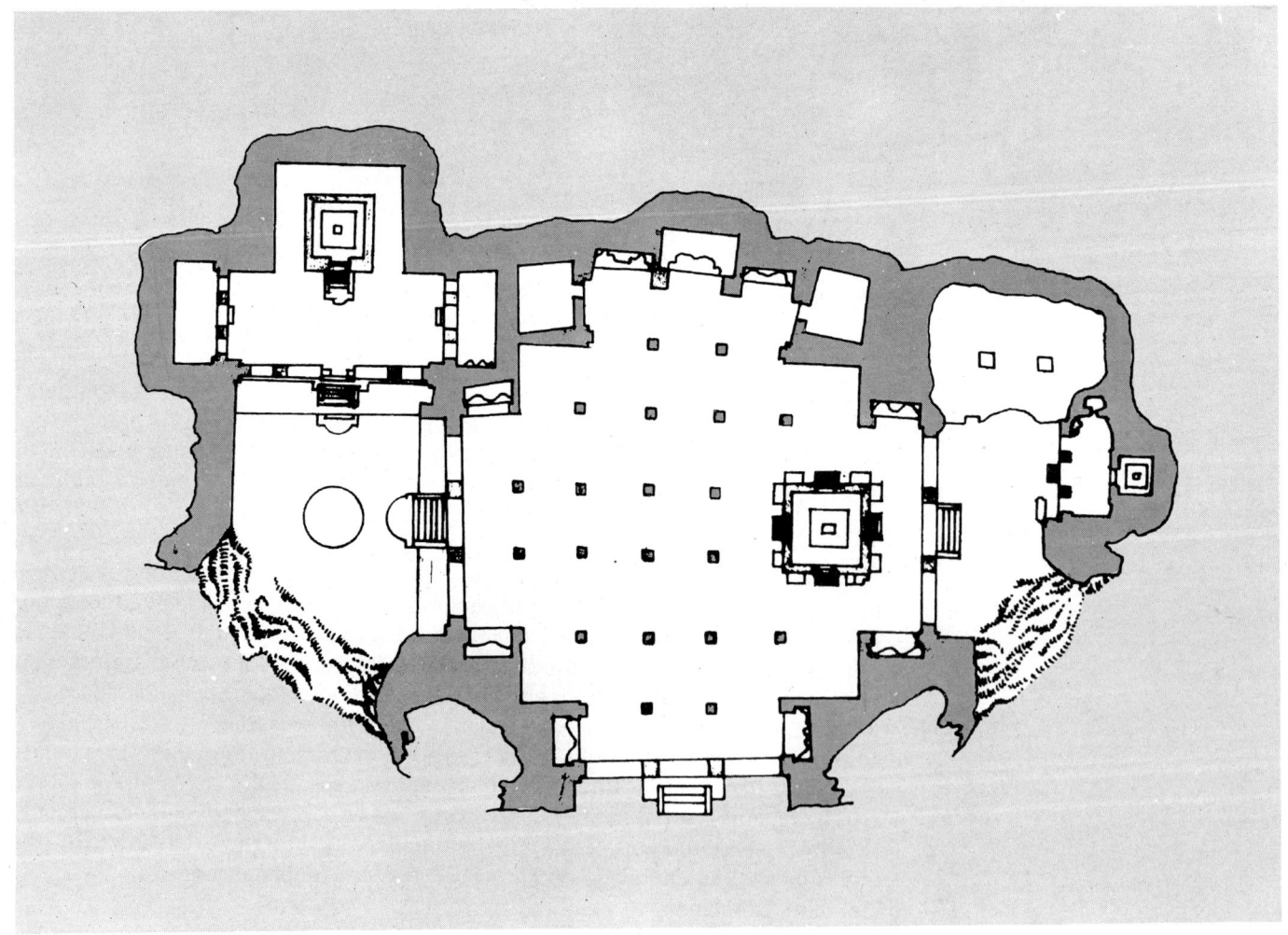

Fig 12.11 The gigantic sculpture of the trimurti at Elephanta that is one of the great triumphs of the Indian sculptor's art

The Great Kailasa at Ellora

The scene now shifts to Elapura (now Ellora) near the Rashtrakuta capital of Kalkhed and only sixty miles from Ajanta. In the early phase, caves that were mere variations of those at Badami or Elephanta, were added to the existing group of Buddhist caves. Architectonic power came into its own when the unrelenting rock cutter picked up once again the strains of the art of Pallava craftsmen of the *rathas* of Mahabalipuram. The fullness of their dreams found expression under the patronage of the great Rashtrakuta king, Krishna I, when they contemplated the creation of the famous Kailasa temple at Ellora.

The Kailasa temple (*Fig 12.12*) is a living rock replica of a structural form, larger than could even have been constructed through the stone masonry technology of the eighth century. Once the idea of carving out a temple equal in area to the Parthenon and one and half times its height had been conceived, its production as well as evolution of design "became a matter of time, patience and skilled labour." That they were setting out on the herculean task of excavating some twenty million tonnes of rock, did not seem to deter the Indian stone carvers.

Their first step, though simple but laborious was to cut three massive trenches to isolate an "island" of rock over 200 ft (65.60 m) long and 100 ft (32.80 m) high from the oftworked cliff-like formations of Ellora. Beginning from the top, the rear mass of the island was gradually fashioned into the shape of a *vimana* to crown the main cell. Each portion of the carved rock was painstakingly finished and polished before proceeding downwards, to avoid erecting cumbersome scaffolding at a later stage.

The flat-roofed *mandapa* in front of the sanctum, seemingly supported on 16 columns in groups of four, the five shrines which surrounded the cella, and a pavilion housing a larger than life sacred Nandi bull, rest on a 25 ft (7.6 m) high plinth (*Fig 12.13*). Such a massive stylobate was obviously meant to lift the entire composition out of the yawning pit around. The Nandi pavilion situated away from the main body of the *stambhas* reminiscent of the Asokan pillars in front of Buddhist *chaitya* hall of Karli are planted on either side of it. The fashioning of the front screen of natural rock into an entrance gateway completed the essentials of the main temple (*Fig 12.14*).

*▷ Facing page
Fig 12.12 Fifty million tonnes of rock were won with chisel and wedge to create the trench out of which loomed the once blazing white AD 800 temple of Kailasa at Ellora*

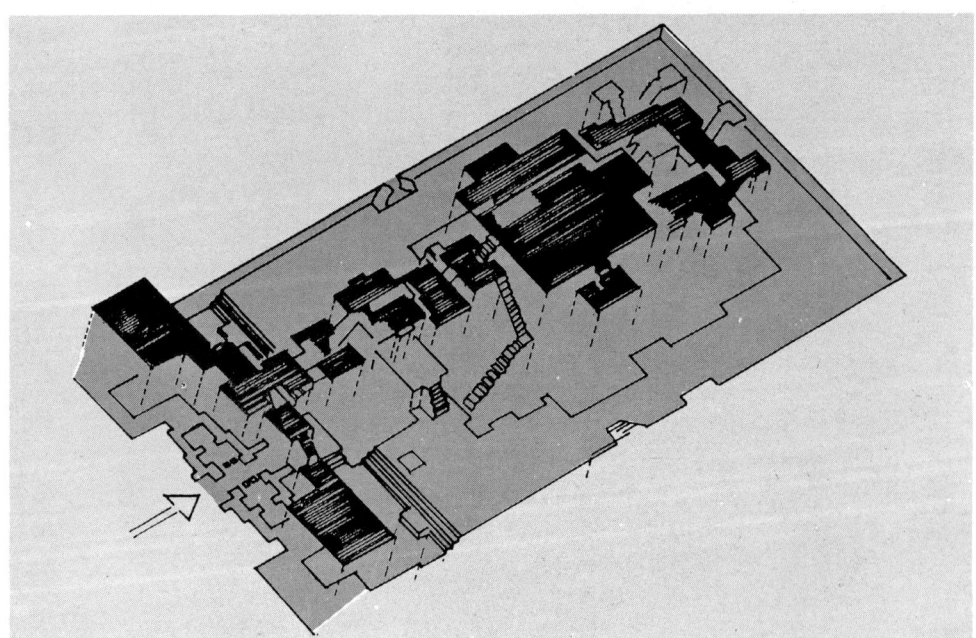

Fig 12.13 Platforms were formed at various levels to create a base for the sanctum, the mandapa and the Nandi pavilion

To the untiring rock cutter, however, the craggy surfaces that had been exposed from the surrounding cliffs, were a challenge in themselves. Into this scarp he tunnelled a gallery for circumambulating the temple and providing approaches to two subsidiary cave chapels carved out of the two sides. These in turn were at one time linked to the main temple by bridges which have now vanished.

In spite of its gigantic scale, the temple seemed as lost to the builders as it does to the visitor today, against the immense backdrop of overhanging cliffs of the same texture and hue as the shrine. In an effort to highlight its profile from

the drabness of the surrounding pit, the stucco worker covered the entire surface of the Kailasa temple with a white gesso. When complete, it shone brightly, like the snowcapped peak of the Mount Kailasa in the Himalayas, the mythological earthly abode of the God Shiva. The sculptures on the plinth were at one time coated with a think layer of polychromed plaster; no wonder the Muslim invaders of the sixteenth century who saw it in its blazing glory referred to the temple as the "Rang Mahal" (the coloured palace).

That such a massive achievement as the Kailasa could even be repeated, was inconceivable even to the legendary architect responsible for its completion. He, it is said, admitted that "it could only be by magic that I could have constructed even this one." That magic truly was never repeated; the Kailasa was to be the great and glorious swan-song of rock architecture in India. Although the Jainas who won the patronage of the last Rashtrakuta king also carved out cave temples of their faith at Ellora, including a mini-imitation of the Kailasa, there was nothing to equal the splendour and magnificence of the original.

In the ninth century, temples, apart from serving purely religious functions, were also becoming symbols of temporal power and the focus of urban social life. The rock-cutters' art alas! but its very nature could be practised only in mountainous and difficult rocky sites. Since the more prominent capital cities of feudal times were situated in the plains, the future of the art of rock carving was bleak. In giving to the rock-cut Kailasa the appearance of a splendid structural temple, the stone carver had, in a way, himself put the seal of doom on his own art. In its gigantic proportions alone, the Kailasa was like a gauntlet thrown at the stone mason, challenging him to create, if he could, edifices as lofty as the Kailasa.

Fig 12.14 A view of the great Kailasa, seen against the gigantic man-made cliffside

CHAPTER 13

The Feudal North

Rashtrakuta power located as it was on the Deccan Plateau of Central India, overlooked the southern regions as well as the plains of the North. It may well have subjugated the entire South, had the Rashtrakuta kings resisted the temptation of rising to a national power. Their political ambitions, however, proved larger, and they continued to cast covetous glances at the city of Kanauj in the central plains. Thanks to King Harsha, Kanauj enjoyed the status that Pataliputra had under the Mauryas; the capture of this city was tantamount to paramountcy over northern India.

The Rashtrakutas Look North

In trying to gain their objective, the Rashtrakutas had to contend with the rising power of the Pratiharas in the West and the Palas of Bengal in the East. The Pratiharas (literally "door keepers") were probably a dynasty of palace officials risen to power. Their fighting strength had been mettled by successfully resisting pressures from the Arabs who had by the eighth century established large Muslim colonies in the Sind. The Palas, unknown until recently, had risen to prominence under the leadership of Gopala. King Gopala's armies soon prevailed in the traditionally rich area of Bengal and Bihar, and his kingdom had substantial commercial interests in South-east Asia. It was, however, Gopala's son Dharampala who resolutely pitted Pala power against that of the Pratiharas and the Rashtrakutas for the conquest of Kanauj.

The Legendary City of Kanauj

The three contestants turned out to be evenly matched in their war power. None was able to establish undisputed suzerainty over northern India, or even Kanauj. Paradoxically, in spite of—or more probably because of—being the cynosure of all eyes, including the Turks in the tenth century, the otherwise legendary city of Kanauj has not acquired any prominence in the known history of Indian architecture. It was so frequently sacked and demolished by warring armies that it is difficult to trace the remains even of any religious or secular ancient building of any consequence.

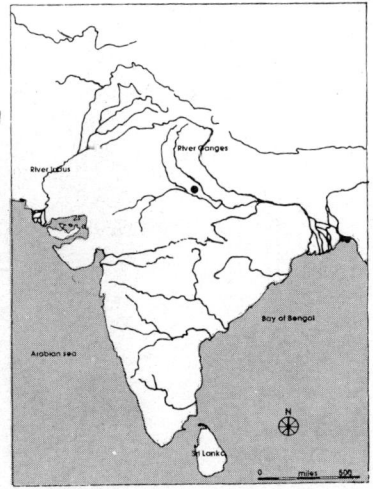

Religious Fervour and Feudalism

From the time of the sack of Kanauj by the Turks, to the rise of Delhi 300 years later, no city enjoyed the status of being the political hub of the North. Instead, a number of provincial capitals built to celebrate the short-lived splendour of their feudal overlords sprouted up all over the country (*Fig 13.01*). Apart from the inevitable fine differences in detail, similar social and political conditions seem to have prevailed in the various mini kingdoms. Under the circumstances, it is confusing and often even redundant to describe the architecture of the north in the middle ages in relationship to the ever changing temporal powers. Rather, it is more pertinent to understand that the euphemistic "religious fervour of the times" that gave rise no doubt to the building of thousands of temples all over the countryside was the result of an alliance, or almost conspiracy, between the ruling and priestly classes. The latter diverted the attention of the ever suffering commoner from his daily woes to promises of a blissful existence through the worship of the expanding Hindu pantheon. In return, the ruling Kshatriya class paid exemplary tribute to the Brahmin, which really amounted to footing the bill for the performance of elaborate rituals and the building of great and many temples.

Fig 13.01 Map showing important sites of the medieval kingdoms of India

Temple Building—the Only Permanent Art

An architecturally consequential part of the deal seems to have been to agree to erect only religious edifices in permanent materials like stone. Secular architecture, including palaces for royalty, were assumed to be of a transitory second class nature and built in perishable materials like brick and timber. Thus, little secular architecture of the feudal period of any consequence has survived intact to date. And though, as we shall see, rules were propounded even for the "third class" architecture of houses for the common people, these were obviously so humbly built that not a trace is to be found of them except, of course, in the dictates of the Brahmins as contained in the ancient theological treatises on Hindu architecture described later. The perceivable history of Indian architecture of the middle ages continues, therefore, to be a story of the evolution of the art of temple building.

The conventional method of tracing the development of temple architecture in the North, has been to "treat its various phases not dynastically, but geographically so that each may be described according to the region in which it flourished." This, however, is to miss the point that throughout the development of the temple in the North, "there is an undercurrent of thought ... which indicates that all these examples belong to the same wide movement." No doubt, craftsmen working in a particular region evolved a distinct "school" of their own, yet they were often influenced by the ideas developed in other regions.

The aim of this study being to trace the evolution of design of major forms of Indian architecture, it is often desirable, or even necessary, to take liberties with the strictly regional classifications. The germ of an artistic vision often blossomed to perfection after crossbreeding with the peculiar genius of other regions. For our purposes, therefore, it is inevitable that we hop from place to place and style to style to see how the shape of the *shikhara*, or the *vimana* or the *mandapa* or the *garbhagriha* was moulded into endless number of variations by the ingenuity of the various schools of craftsmen working in the far-flung corners of northern India.

Typical shikhara and the vimana

Temples of Osian

To pick up the strains once again, one finds that the earliest subsistent link in the evolution of the *shikhara* (the birth of which during the heyday of the Gupta empire has already been described in detail) is to be found in the village of Osian near modern Jodhpur in Rajasthan. In the eighth century, it was probably the capital of the Pratiharas and "is an example not uncommon in India of a considerable city, whose substance has departed and only the spirit remains"; the spirit being a handful of small but elegant temples crafted not in the familiar brick masonry of the Guptas, but in everlasting stone.

The *shikhara* that had become the crowning feature of the sixth century Hindu shrine of Deogarh, was now adapted into the design of a full-fledged Osian temple such as the one classified as "Temple No. 2." To the cella and *shikhara* of their prototype, the builders of Osian added a *mandapa* or open assembly hall supported on columns (*Fig 13.02*). The edifice is raised on a high plinth, the outer surface adorned with the rich plasticity of the sculpture of the Gupta period. In grafting the *shikhara* into a comprehensive temple design the false porticos of the

Fig 13.02 The Deogarh type shikhara is adapted into the so-called temple no. 2 at Osian, consisting of a mandapa on slender columns and with a sloping stone bench

Fig 13.03 A more perfect example of the Osian group of temple is the Surya temple of AD 900

Deogarh example were found to be redundant. This omission of a classical idiom is more than amply compensated by the retention of a few rich folk elements. The sloping stone seat of the Lad Kahana type now exquisitely decorated, surrounds the *mandapa*. Also, the plinth of the temple is stopped short of the end portico columns, which are carried right down to the natural ground adding to their slenderness (*Fig 13.03*). Perhaps unintentionally, the builders introduced a flair that matured into a classic form in a subsequent temple at Osian, that of Surya (the Sun god).

In the Surya temple (*Fig 13.04*) the *shikhara*, instead of being a slender tapering form, acquires more body and rises up in a single, more confident curve towards a disc-like *kalasa* stone capping the apex. The *mandapa* too is equally dignified. Shedding some of its folk trappings, it emerges as a rectangular hall with its flat roof held up over rows of stately square columns rising up from a substantial plinth. The slenderness of the columns of the entrance portico, which rise straight up from ground level, is emphasized by vertical flutings on their circular shafts. The temple was at one time set within a rectangular precinct defined by a cloister, at each corner of which was located a subsidiary shrine.

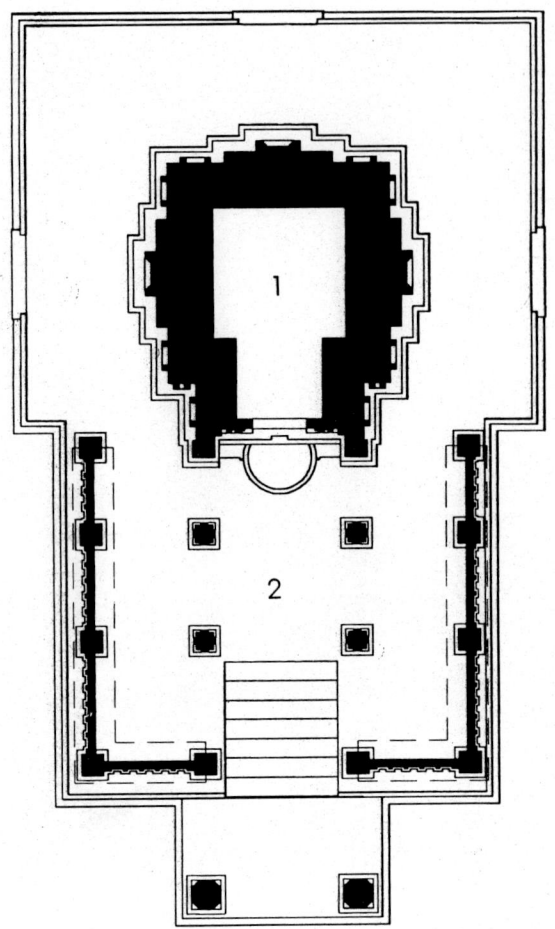

Fig 13.04 Plan of the temple of Surya at Osian 1 Garbhagriha and 2 Mandapa

Hindu Temples in the Valley of Kashmir

Development of the Hindu temple along parallel lines was meanwhile continuing in other urban centres of the plains. Let us, however, also have a look at the contemporary styles developing in the more isolated hill states such as Nepal, Kashmir and Garhwal. The distinctive architecture of these regions was more suited to their geographic and climatic conditions, and even absorbed traits from the styles of neighbouring nations. The most flourishing of the hill states that falls into this purview is that of the fabled valley of Kashmir. In the sixth century AD, the Buddhists who had fled from the wrath of the Huns, had built a number of monasteries in Kashmir. The climate being none too kind to brick masonry, not one of these has survived in a recognizable architectural form. In the eighth century, Kashmir was under the rule of King Lalitaditya who had developed it into a prosperous and virtually self-sufficient state. The ambitions of the Hindu King Lalitaditya, however, were not entirely restricted to the valley. He made several successful forays south into the plains, and was also partly responsible for warding off the threat posed by Arab expansionism in the Sind. By opening up channels with the plains, Lalitaditya exposed the valley to the influence of Hinduism, which ultimately prevailed over Buddhism. The craftsmen of Kashmir, though, were not a part of the stream of architectural development in the plains. With their background of Buddhist artistic traditions, they evolved a style of Hindu temple architecture, with a distinct flavour of its own.

*Fig 13.05 The Hari-Hara
Temple at Osian*

The earliest known Hindu temple, that of Shankaracharya (*Fig 13.06*), is located on a hill top overlooking the famous Dal Lake of modern Srinagar. Built in the eighth century on an octagonal platform, it is approached by an imposing flight of steps and has an unusual plan. Though from outside the shrine appears to be the conventional square in plan (*Fig 13.07*) with recessed chases running along its periphery, it encloses in fact a circular cella, echoing the shape of the rounded Shiva-lingam installed in it. What distinguishes the temple visually from

Fig 13.06 The Sankaracharya at Srinagar like many other living shrines has suffered due to various unplanned architectural modifications. At one time, it must have been an imposing edifice

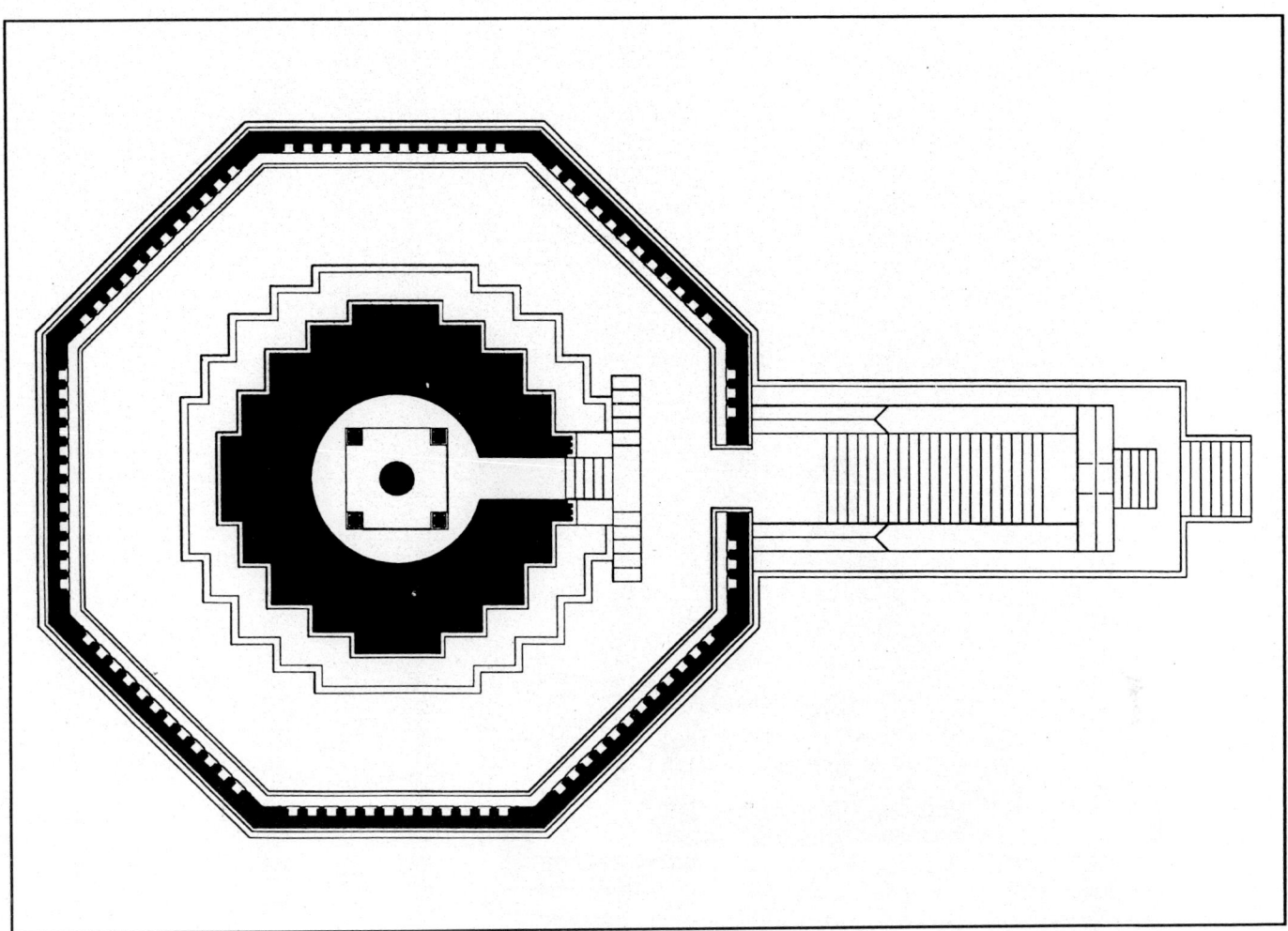

Fig 13.07 Plan of the Sankaracharya temple at Srinagar

its prototype of the plains though, is the entrance portico. This is in the form of a stone abutment to the entrance doorway of the cella and contains a triple cusped trefoil arch, set inside a high pitched triangulated pediment. The trefoil arch is obviously derived from the Buddhist *chaitya* arch, with modifications effected by the Greco-Roman traditions of Gandhara craftsmen. The triangular pediment, however, could have been derived either from the gable end of the pitched timber roofs common in the hill country around the valley, or is an ungainly steep derivation of the classical Greek pediment, well known to the Gandhara builders. The tower that crowns the cella, instead of being the curvilinear *shikhara* is a plain unornamented pyramid. Its smooth sloping surfaces do not make for a very attractive form but are eminently suited to ward off the winter snows, just like the timber steeples of the shrines of the hillside hamlets.

A distinguished, but modest shrine like the Shankaracharya was, however, hardly likely to satisfy the growing ambitions of the Kashmir kings. Commanded to built more elaborate and glorious shrines, the local architects instinctively turned to Buddhist prototypes for inspiration. This is evident from the famous Sun Temple of Martand (*Fig 13.09*) and the temple of Avantipur which are laid out much like the great stupa courts of Takht-e-Bahi and Takht-e-Jamshed, built in the Gandhara country over 400 years ago.

Hill Construction Techniques

A word must be said here about the techniques of construction employed by the hill craftsmen. The eighth century builders of Kashmir were way ahead of their contemporaries of the plains and peninsular India. While the Hindu mason of the plains was content to built his temples with stone masonry laid dry, the walls of the temples of Kashmir were constructed of evenly dressed ashlar masonry, carefully jointed together either with lime mortar or even steel dowels; techniques that may well have been learnt from the craftsmen of neighbouring western regions. However, as we have seen, even these refined techniques could not stand up to the rigorous climate of the region, and only a few of the vast number of temples that must have been built have survived (*Figs 13.08, 13.09*).

Fig 13.08 Plan and elevation of a typical small countryside Hindu shrine showing the hill construction techniques typical of Kashmir

Lalitaditya's successor, King Avantivarman, heralded the beginning of his rule by laying out a new capital city—that of Avantipur, overlooking a bend in the Jhelum, 18 miles south-east of modern Srinagar. Its chief attraction must have been the Avanti Swami Temple which, if anything, is in an even worse state of preservation than its prototype, the Sun temple at Martand.

Fig 13.09 Ruins of the great
eighth century Sun temple at
Martand, Kashmir

At one time, however, the Avanti Swami temple (*Fig 13.10*) possessed the necessary wherewithal for the performance of elaborate Hindu rituals, including a ceremonial bathing tank, a *kirti stambha* or heraldic pillar and subsidiary shrines in each corner of the court (*Fig 13.11*). In its heyday it must have presented a stately picture of the classical *pancharatana* type of temple plan. Its architectural character, however, had not evolved much beyond that of Martand.

Fig 13.10 Plan of the ninth century Avanti Swami temple near Srinagar. 1 Sanctum, 2 Peristyle and 3 Entrance portico

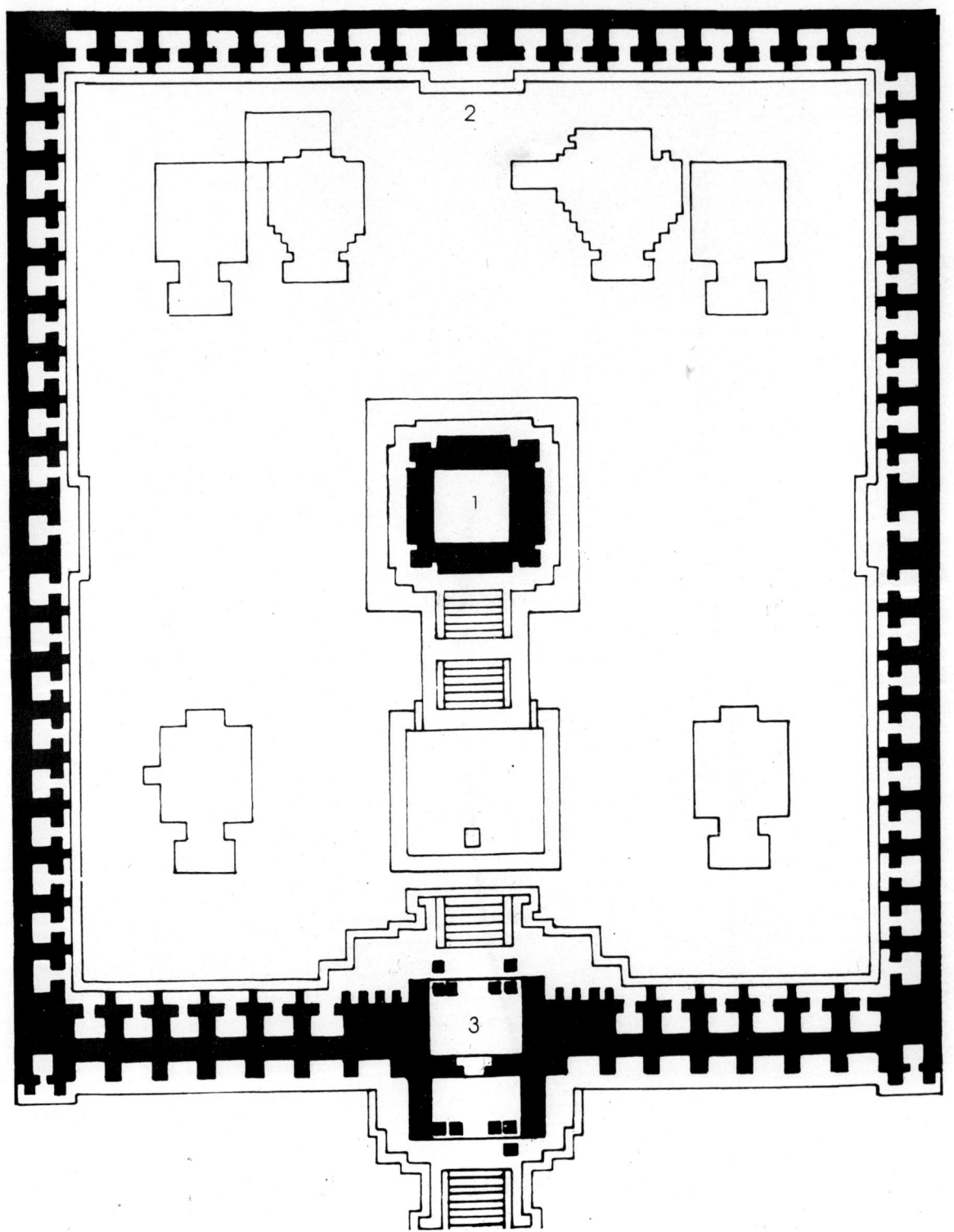

Fig 13.11 A conjectural restoration of the Avanti Swami temple as it must have been during the regime of King Lalitaditya in Kashmir

The Sun Temple at Martand

Martand, victim both of the harsh climate as well as the ultimate decline of Hindu power in Kashmir, today is a heap of uncared for ruins (*Fig 13.12*). It is nevertheless evident that in its design, the plan of the Buddhist monastery has been adapted to create a new form for the Kashmiri Hindu temple (*Fig 13.13*). The central place of the sacred stupa is taken up by a shrine of the Shankaracharya type. The form of the trefoil arch abutment that was the entrance to the Shankaracharya has been added on all the four sides, and the temple is set on a high square platform. The perfectly symmetrical design also reminds one of the temple of Deogarh. What were cells for the residence of monks in the Buddhist monastery are transmuted into so many niches, adorned with quasi-Doric columns and filled with images of the unending pantheon of Hindu deities. The entrance pylon to the rectangular courtyard defined by the niches is, in its architectonics, a replica of the central shrine.

Fig 13.12 The impressive remains of the great Sun temple at Martand

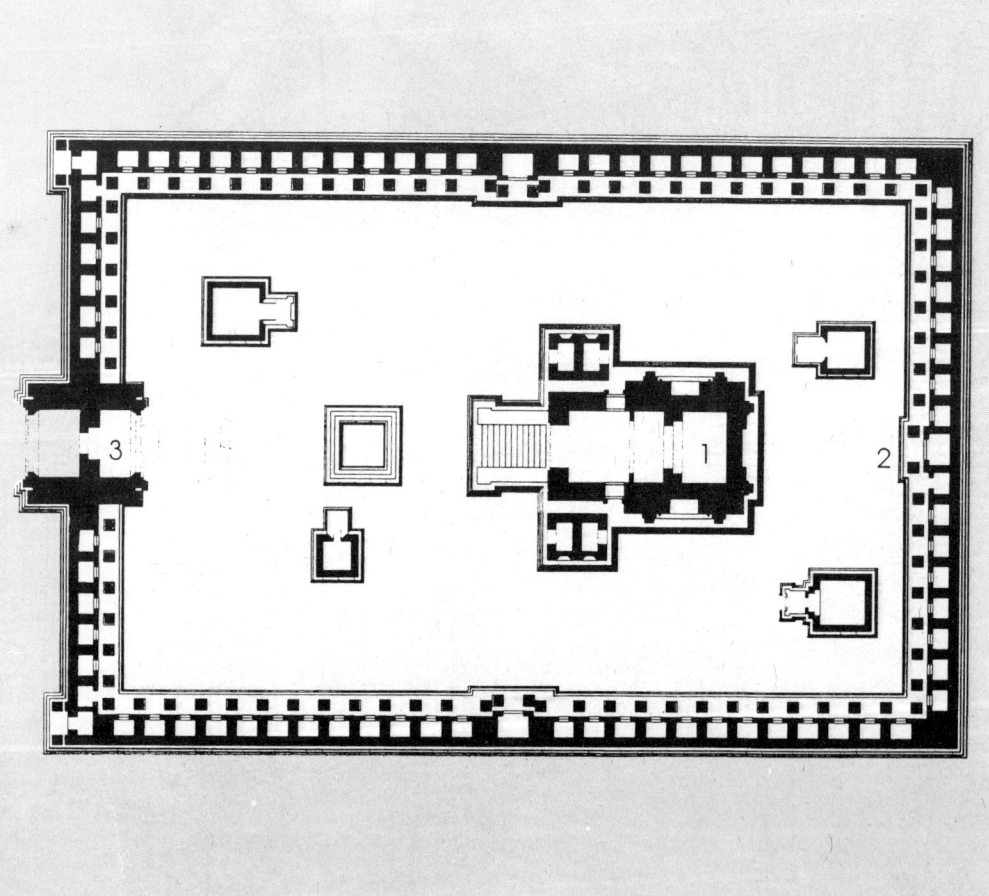

Fig 13.13 Plan of the great Sun temple at Martand. 1 Sanctum, 2 Peristyle and 3 Entrance portico

CHAPTER 14

Rajput "Vanshas" and the Kalinga Kings

Leaving Kashmir, which rose again to architectural prominence only under the Great Moghuls more than 500 years later, we come down once again o the plains. The Hindu mason, in spite of his rather rudimentary structural techniques, was busy erecting temples for his proliferating overlords. The fluxionary state of eighth century feudal India, gave rise to many centres of feverish building activity. The most prominent of these were located in the central region of the hills surrounding Gwalior, the western plains of Gujarat and Kathiawar, the eastern coast of Orissa and in the desert strongholds of Rajasthan in the north-west.

Teli-ka-Mandir at Gwalior

While Gujarat, as we shall see later, took its cue from the neighbouring style of the temples of Osian, the so-called Teli-ka-Mandir (literally 'oil man's temple') within the hilltop fort of Gwalior is in a class by itself. Local legend attributes its peculiar name to the fact that it was built from the donations given by an oil merchant. Judging from its style, it is more probable that the name is a distortion of Telengana (modern Andhra) suggesting that it was designed by an architect from that region. In fact, it was the last grand attempt at adopting the Buddhist *chaitya*

Fig 14.01 Plan of the Teli-ka-Mandir at Gwalior

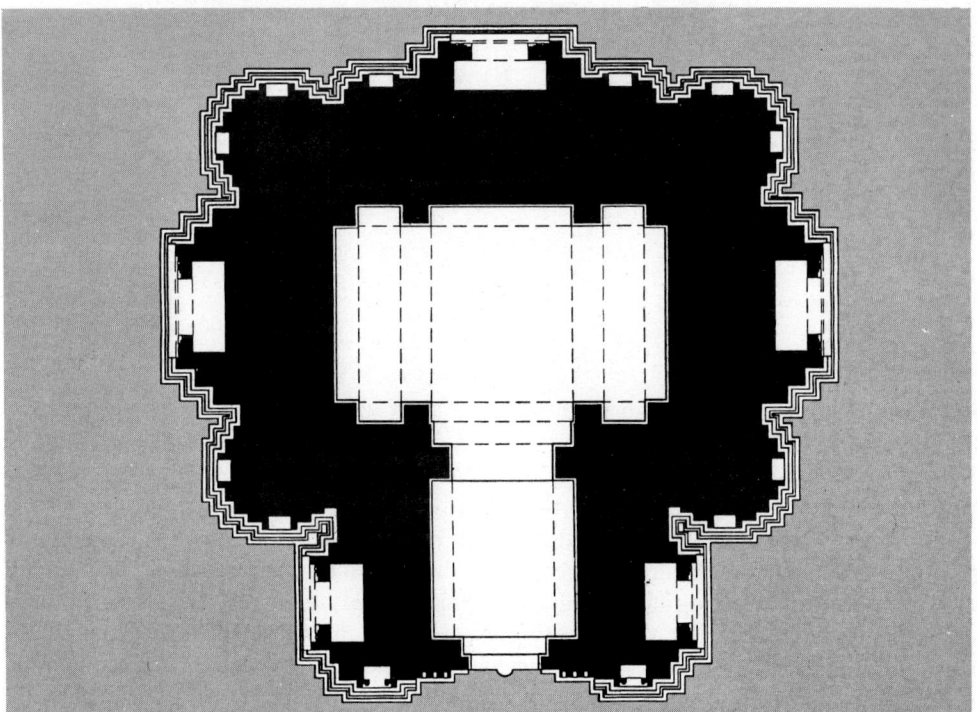

hall roof over a rectangular plan (*Fig 14.01*) as a model for a Hindu shrine. One will recall that this was one of the seven forms of the *rathas* carved out at Mahabalipuram in the South. The pile of masonry of the Teli-ka-Mandir (*Fig 14.02*) with richly sculptured horizontal friezes, towers 80 ft (24.3 m) into the sky. Half way up, the rectangular prism of the base becomes a pyramid, crowned by the hump of the barrel-vaulted roof form so popular with the Buddhist architect. Though a form discarded by the northern Hindu architect, a variation of it, as we shall see, was destined to dot the southern skyline in the form of massive ceremonial gateways of its myriads of temples.

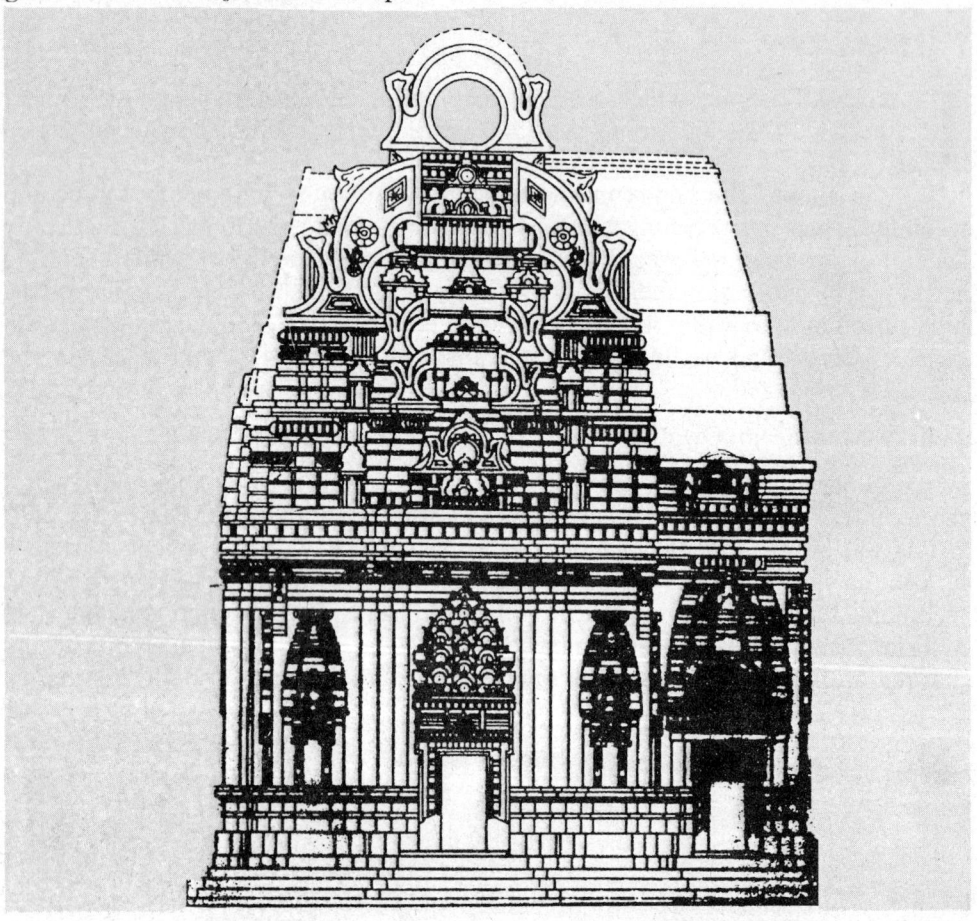

Fig 14.02 The 24 metre high tower of the Teli-ka-Mandir (AD 800). One of the few examples of a northern Hindu temple adorned with a Buddhist chaitya hall type roof

The Vanshas

The desert region of Rajasthan in the middle ages was peopled by martial clans, the forefathers of the modern Rajput. The foremost of these were the Parihars, Chauhans, Parmars and Solankis. They were the descendants of warlike tribes who were forced to proliferate into the desert region of India, under the pressure of the Hun invasions of the north-west region in the fourth century AD. They had by now persuaded the Brahmin keepers of the Puranas (traditional records and stories of Hindu genealogies) to accord them not only Kshatriya warrior status within the caste system, but also genealogies, claiming their descent from legendary solar and lunar races, the popular Suryavanshi and Chandravanshi. The social status even of the modern proud Rajput is based on his belonging to either of these major 'vanshas'.

The Suraj Kund, Delhi

The Tomars, an offshoot of the Rajput Chauhans, had in AD 736 founded the city of Dhillika (modern Delhi) on the banks of the Jamuna river. It is said to have contained more than 20 temples of the Hindu and Jaina faith. Subsequently, in the eleventh century, building material salvaged from most of these was absorbed into the Muslim architecture of Delhi. That the Tomar craftsmen had reached a high level of temple building is testified even by the ruinous remains of the so-called Suraj Kund in the south of Delhi (*Fig 14.03*). This at one time must have been a massive temple, probably dedicated to the Sun God, overlooking a large circular ritual bathing tank. Today only the tank remains, that too in a ruinous state, the main shrine having completely vanished. A similar fate seems to have overtaken the creation of the Parmar kings in the central Indian region of Malwa, which too was ultimately devastated by the Muslims.

Fig 14.03 Ruins of the platform and tank of the eighth century Suraj Kund near Delhi, where at one time stood a temple dedicated to the Sun God

The Eternally Holy Kalinga

Of the other states, the land of Kalinga (modern Orissa) the conquest of which had been the turning point of Asoka's career, was coming into the limelight once again. It was not only one of the major springboards for the spread of Indian culture to the far east, but was also emerging as a flourishing centre of temple building activity.

The temporal history of this region, like that of most of India at the time is vague and obscure. The little that is known suggests that Kalinga, in the true feudal tradition, was ruled by a succession of kings, who sought personal glorification through the building of temples rather than empires or cities. Orissa is yet another of the many examples of the crafty collusion whereby the king and priest ruled over a "people (that) were quite content to live in hovels and mean habitations themselves, but for the houses of the gods, they would not stint either expenditure or labour." The cities of Kalinga then, are not surprisingly more renowned for their intensely devotional fervour and great temples than their civil architecture, devoid as they are of any semblance of town planning. Over a period of seven hundred years, the holiest of the holies, the city of Bhubaneshwar, came to acquire some seven thousand temples and little else.

Vaital Deul at Bhubaneshwar

Early experiments in temple building were made with forms derived from Buddhist architecture, as in the seventh century—Vaital Deul temple at Bhubaneshwar (*Fig 14.04*). The 35 ft (10.6 m) high 'deul' or shrine of this temple is in fact a

Fig 14.04 Another attempt to adopt the Buddhist medium to a Hindu temple—the Vaital Deul temple at Bhubaneswar (AD 850)

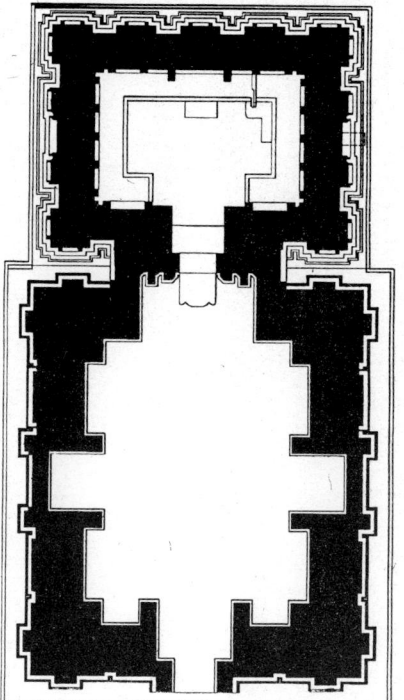

Fig 14.05 The Vaital Deul at Bhubaneswar showing shikharas, at the four corners. Plan of the Deul below

Fig 14.06 The typical "shoulder-shaped" Orissa shikhara

refined version of the Teli-ka-Mandir at Gwalior described earlier. Like the latter, this too is surmounted by a roof-form of unmistakable Buddhist ancestry, the familiar *chaitya* hall vault. The similarity, however, ends here since the Orissa craftsman grafted many original elements into his design. For example, he chose to leave unadorned the outer surface of vault and instead planted three *kalasa* type finals, generally found over a *shikhara*, on the ridge of the lower vault. What is more, the form of the familiar North-Indian *shikhara* is attached to the corners of the otherwise undistinguished rectangular flat-roofed *mandapa* in front (*Fig 14.05*). The rather incongruous introductions of the *shikhara* even as a subsequent element into the design, indicated the general trend of the Orissa builder towards favouring the *shikhara* as a more acceptable final for the Hindu temple. However, ignoring the pronounced elliptic curve of the *shikhara* of the Gupta period for ease of construction, the Orissa craftsman chose to interpret the *shikhara* as an almost perpendicular prismatic tower. The vertical profile of the Orissa *shikhara* converged only near the apex, towards the circular *kalasa* or crowning stone at the top. This so-called 'shoulder type' spire (*Fig 14.06*) became in due course the distinguished feature of Orissa temples. Simple shrines consisting of a small 'Sri Mandir', or 'deul' as the main cella is called, crowned by the typical Orissa are in the *shikhara* style, grouped all over the Orissa countryside much like wayside Buddhist stupas in Asoka's time.

*Fig 14.06 (a) The typical
Orissa shikhara, the Linga Raja
temple, Bhubaneswar*

A Mandapa for the Deul

Inevitably, the need was soon felt for attaching a *mandapa* or covered hall to single-roomed shrines wherein worshippers could congregate and sing devotional hymns to the enshrined deity. For this purpose, often an existing simple shrine would be expanded. In the eighth century temple of Parasurameswar (*Fig 14.07*), the earliest known example of such a modification, an extremely ponderous low-slung structure, was attached to the Sri Mandir of an earlier period. Rectangular in plan (*Fig 14.08*), the central aisle of the *mandapa* along its longer side is flat-roofed. The side aisles are covered with massive stone slabs laid to slope, leaving between the two roofs a narrow horizontal opening forming a sort of clerestorey lighting of the interior. The walls are made of cumbersome, almost cyclopean blocks of stone, the opening within being more appropriate for pygmies. In structural techniques the entire effort is not much advanced from the rather primitive achievements of the seventh century temple builders of Aihole.

Fig 14.07 The Parasurameswara temple showing shrine and mandapa, AD 800

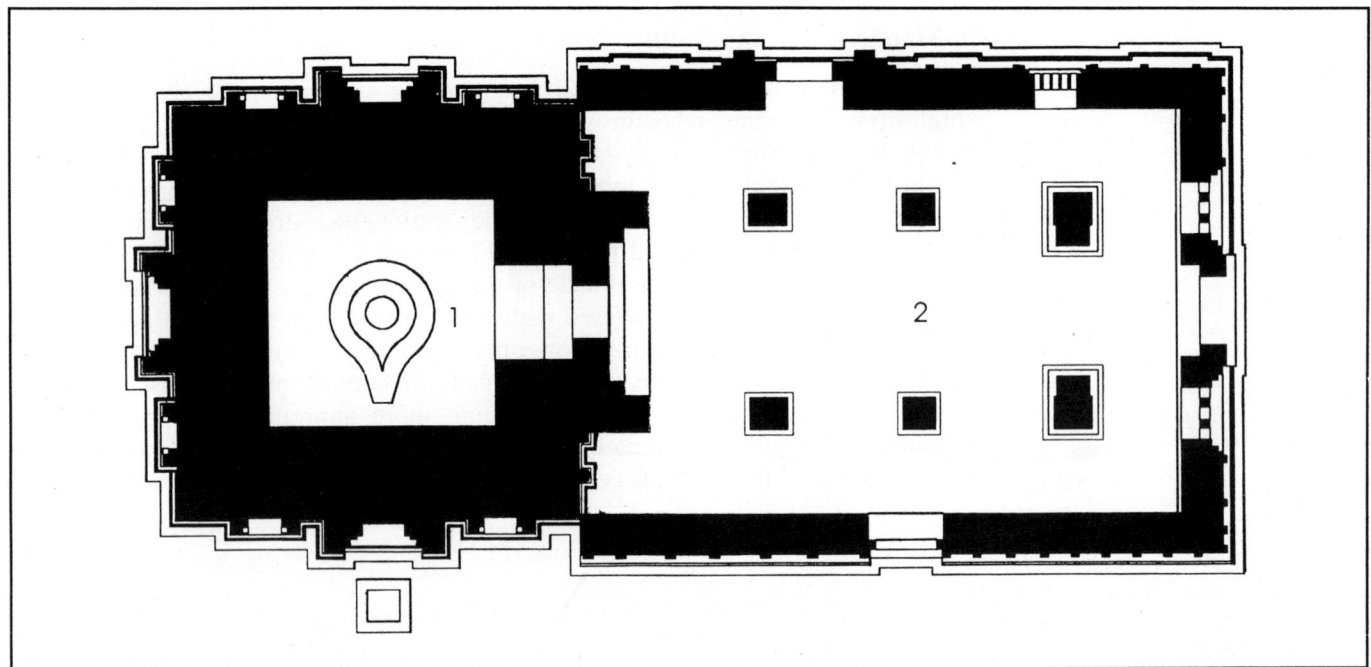

Fig 14.08 *Plan of the Parasurameswara temple. 1 Garbhagriha, and 2 Mandapa*

The Little Gem of Mukteswar

Once an enlarged temple including the *mandapa* became customary, a more elegant edifice such as the temple of Mukteswar (*Fig 14.09*), also at Bhubaneshwar emerged. So far in Orissan architecture the *mandapa* had been treated merely as a room, somehow or the other appended to the door of the main shrine. We have seen how at Aihole, Osian and the early Orissan temples, it remained rather incongruous and uncertain both in design and construction. Now for the first time, the *mandapa* received due attention from its designer. Almost instinctively, it would seem the Hindu architect's first act was to make it square rather than rectangular in plan. He then proceeded to roof it with a tapering form, but realizing the subservience of the *mandapa* to the 'deul', he designed it as a shallow pyramid substantially lower than the *shikhara* tower. It was constructed as a series of corbelled courses of stone, which appear externally as cornices punctuated by small triangular uprights (*Fig 14.10*). In doing so, the craftsmen as ever, were drawing their inspiration from rural forms. The uprights are nothing but a geometric crystallization of the bamboo purlins of the thatch original. Each cornice seems to sail over the lower one; an illusion created by a marked and well shaded gap between two consecutive courses. The inspired hand that devised the form of the *mandapa* also gave clear definition to the temple tower. Each horizontal course of the shoulder-type *shikhara* is well modulated and defined. Entrance to the temple compound is gained through a kind of *torana*, consisting of two vertical pillars spanned by a corbelled though semicircular richly sculptured arch, an elegant element that does not appear again in the temples of Orissa.

The precincts of the Mukteswar temple are defined by a low parapet adorned with lands of richly sculptured friezes. The richness of form, the meticulous planning of both the Vaital Deul and Jagmohan (*Fig 14.11*) under the guiding hand of an anonymous Orissan master architect resulted in a little gem, the like of which was never repeated; not at least in the land of Kalinga.

Fig 14.09 *Plan of the Mukteswara temple at Bhubaneswar. 1 Garbhagriha and 2 Mandapa*

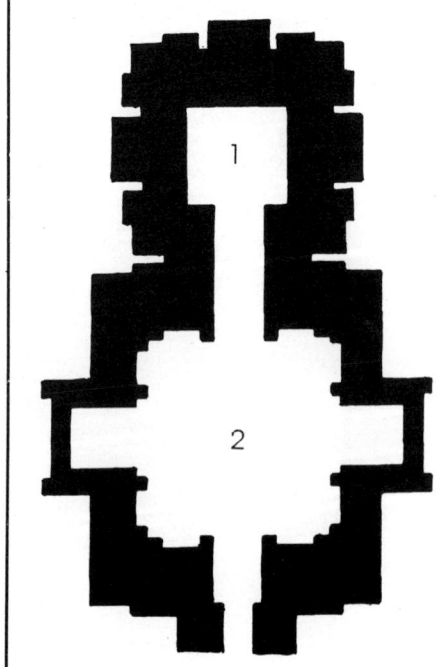

Fig 14.10 The bamboo purlins of the rural form became the triangular stone uprights of the series of deep cornices that are the pyramidical roof of the Orissa mandapa

Fig 14.11 A rare Orissa example of a temple planned and built without subsequent additions—the little gem of Mukteswara at Bhubaneswar (AD 900)

*Fig 14.11 (a) Another view of
Mukteswar Temple, Bhubaneswar*

CHAPTER 15

The Chandelas of Bundelkhand

Almost all other temples of the ninth and tenth century of the Orrisa style display an inherent weakness in establishing an appropriate relationship between the cella and the *mandapas*. The shortcoming may be attributed to the fact that rarely did the Orissa architect get a chance to visualize his temple as a single entity. In his almost two centuries of building in Orissa he was more commissioned to add to existing shrines, rather than to design and execute complete new temples. In the process he seems to have lost the capacity to knit together the elements of his creations into a crystalline architectural statement. This crystallization is to be found in a parallel development which was taking place over 500 miles north-west in the Bundelkhand region of central India. Here the Chandelas, an offshoot of one of the Rajput martial clans, had gained supremacy in about AD 100. Their building craftsmen had gained sufficient practical experience in construction by skilful engineering of large civil works, such as reservoirs and tanks. With equal skill, their colleagues proceeded to embellish the capital city of Khajuraho with temples both of the Hindu and Jain faiths (*Fig 15.01*).

The architects of the Chandelas appear to have been well aware of the visual shortcomings of the temples of Orissa. Even their earliest efforts display a total comprehension of temple design that had eluded the builder of Orissa for a couple of hundred years. Probably the physical proximity of Khajuraho to Orissa and Gujarat where the nucleus of such a design had already emerged also helped the Chandela builders. The largest group of their temples is set in the once desolate and sleepy village of Khajuraho. What at one time was probably the bustling capital of a burgeoning dynasty, is today a cluster of 16 deserted temples. Only in one of these is the deity still worshipped. The rest are a tourist haunt, world famous for their architectural excellence as well as the profusion of erotic sculpture.

The Early Temples of Khajuraho

All the temples of Khajuraho are set on broad high terraces. They were probably "disposed rather unsymmetrically around the borders of an ornamental sheet of water" which today is the rather sterile though well tended gardens of the Archaeological Survey of India. Though the platforms of the temples were sufficiently elevated above the water level, the floor level of most of them is raised by yet another 10 to 12 ft (3 to 3.6 m). This gave the builder the opportunity to erect a stately flight of steps leading from the terrace to the entrance. Over this the plan of the temple is laid out, not as a series of "conjoined buildings but as a compact architectural synthesis" (*Fig 15.02*).

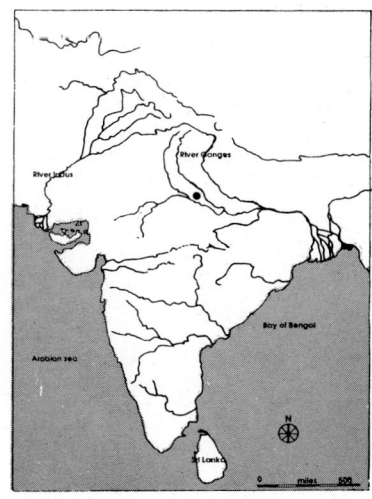

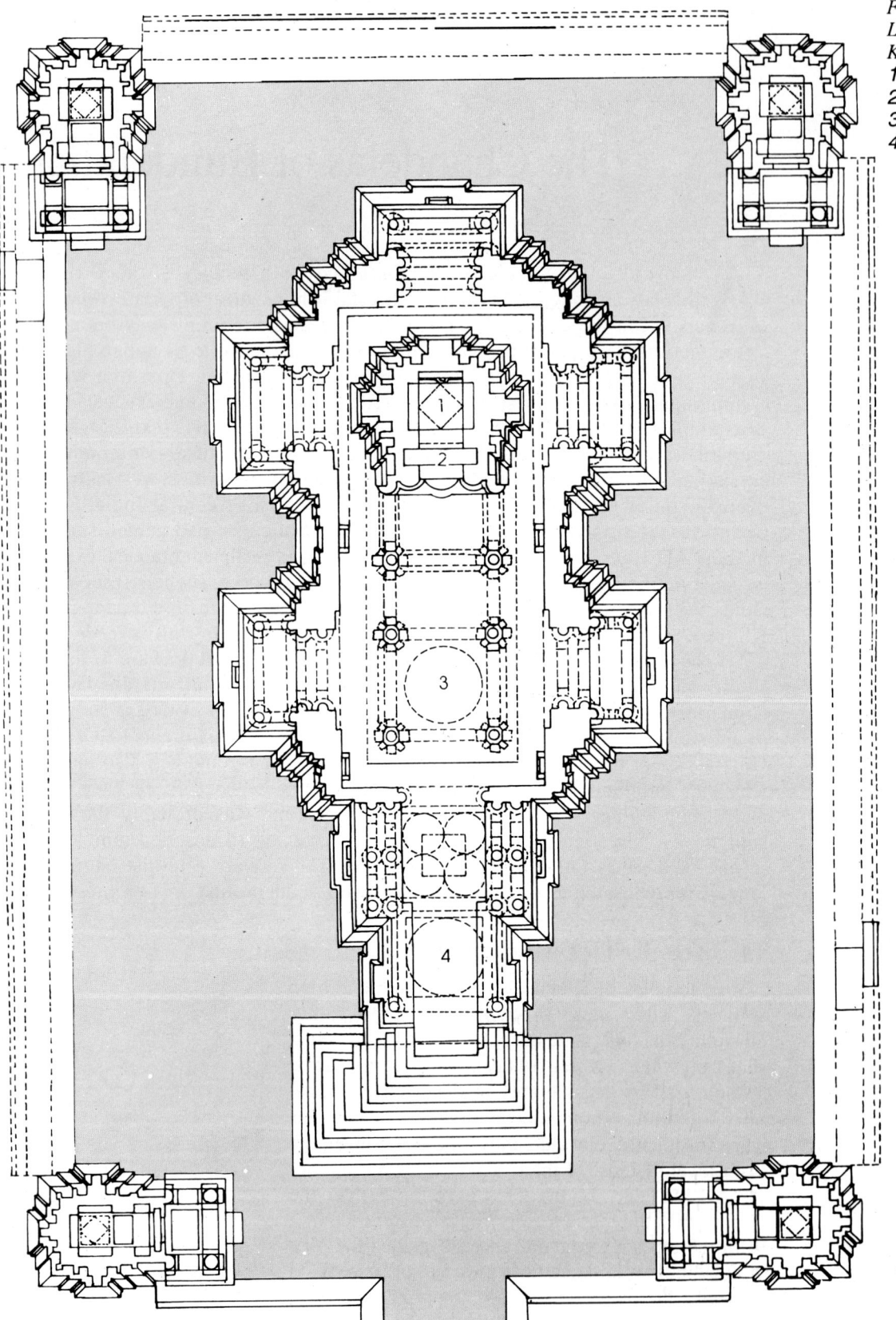

*Fig 15.01 Plan of the
Lakshamana temple at
Khajuraho.
1 Garbhagriha,
2 Antrayla,
3 Mandapa and
4 Ardha mandapa*

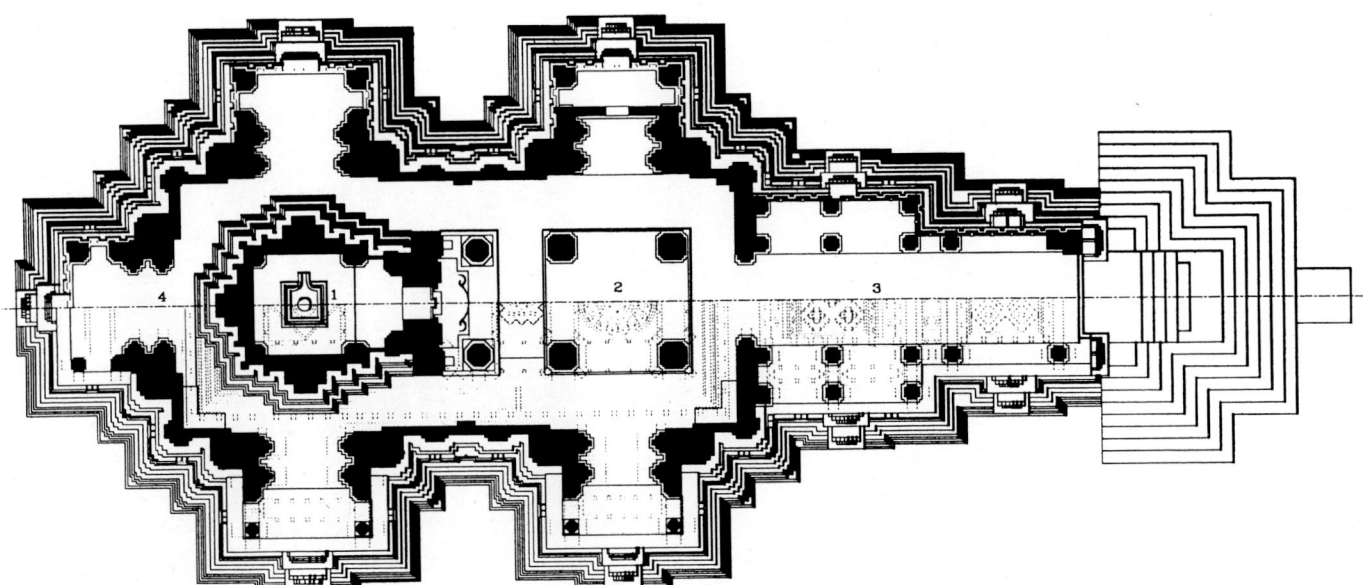

Fig 15.02 Plan of the Vishwanatha temple at Khajuraho. 1 Garbhagriha, 2 Mandapa, 3 Ardha mandapa and 4 Pradakshina

Even more of the comparatively small temples like the Lakshmana (*Fig 15.03*) displays a remarkable unity of composition. The cella, the *mandapa* and the entrance vestibule are parts of a harmonious whole. A narrow pillared aisle with stone benches on either side stretches out to receive the visitor at the top of the flight of steps from the terrace below. Passing through an open *mandapa* supported on four central columns, he approaches the anteroom or *antralaya* immediately behind which lies the dark cubic cella. An elegant skyline of the body of the temple was only a logical outcome of such a precisely delineated plan. The shallow pyramidical roofs over the front compartments ascend gracefully towards the climax of the *shikhara* rising over the cella of the *garbhagriha*.

The Shikharas of Khajuraho

The *shikharas* of Khajuraho are architectural masterpieces in their own right. Gone is the rigid verticality of the Orissa *shikhara*. Instead we find here an ascending parabolic outline rising to the top in one graceful sweep (*Fig 15.04*). Gone too is the tier upon tier of stone courses that tended to compromise the impact of height in the Orissa *shikhara*. On the contrary, and almost defiantly it would seem, the Chandela builders concealed the inevitable horizontal joints of the courses of masonry under clearly defined projecting vertical bands planted in the middle of each of the four faces of the *shikhara*, which rose up towards the *amalka* or capstone, in perfect rhythm with the curvilinear outline of the *shikhara* (*Fig 15.05*). This, however, was only the bare bone of the complex ornamental system that the builders devised for their temple spires.

It was now clear in their minds that the *shikhara*, soaring high into the skies was going to be the crowning glory of their scheme. Would the rather bare parabolic profile of the *shikhara* bands be effective enough for the purpose? The blank surfaces of the vertical bands attached to the body brought up in the Hindu sculptor his inherent sense of "horror vacuum". By now, the Hindu sculptor had acquired the techniques of being able to create visual drama within the parameters of discipline rather than mere gimmickry. He, therefore, devised the simple but appealing idea of ornamenting the body of the monumental form of his *shikhara* with vari-dimensional aspects of its own shape.

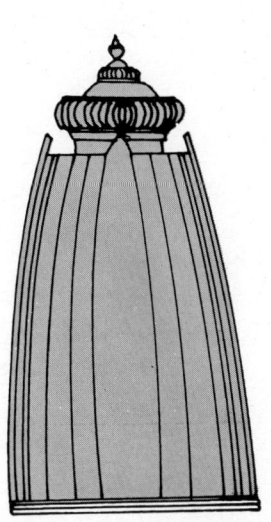

Fig 15.04 An early Khajuraho shikhara

*Fig 15.03 The graceful outline
of the Vishwanatha at
Khajuraho temple (AD 1000)*

Thus, *shikharas* in semi-relief, complete with the *amalka* and *kalasa* (*urusringa*, in Hindu terminology) were applied over the four upper surfaces of the body of the *shikhara* (*Fig 15.05*). This application, he discovered, imparted an extraordinarily smooth sweep to the soaring profile of the spire. Assured by his initial success, the builder had little hesitation in repeating the technique *ad infinitum*. With gay abandon he proceeded to plant *shikhara* over *shikhara* on the main body wherever he appropriately could. Thus, the corners of the base of the tower became like a vertical chain of superimposed full bodied mini-*shikharas*. The body of the tower was filled out and its gracefully ascending profile further enhanced by superimposed gradually ascending *urusringa* upon *urusringa* on the four faces.

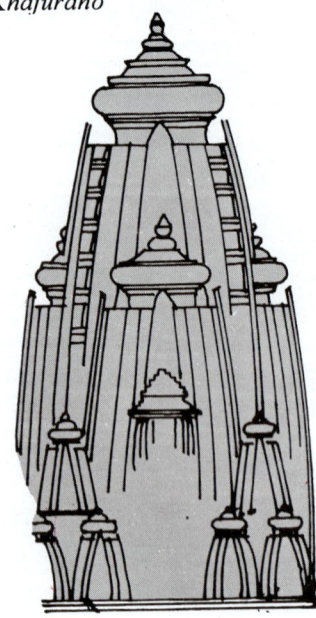

Fig 15.05 Urusringa upon urusringa. The development of the glorious shikharas of Khajuraho

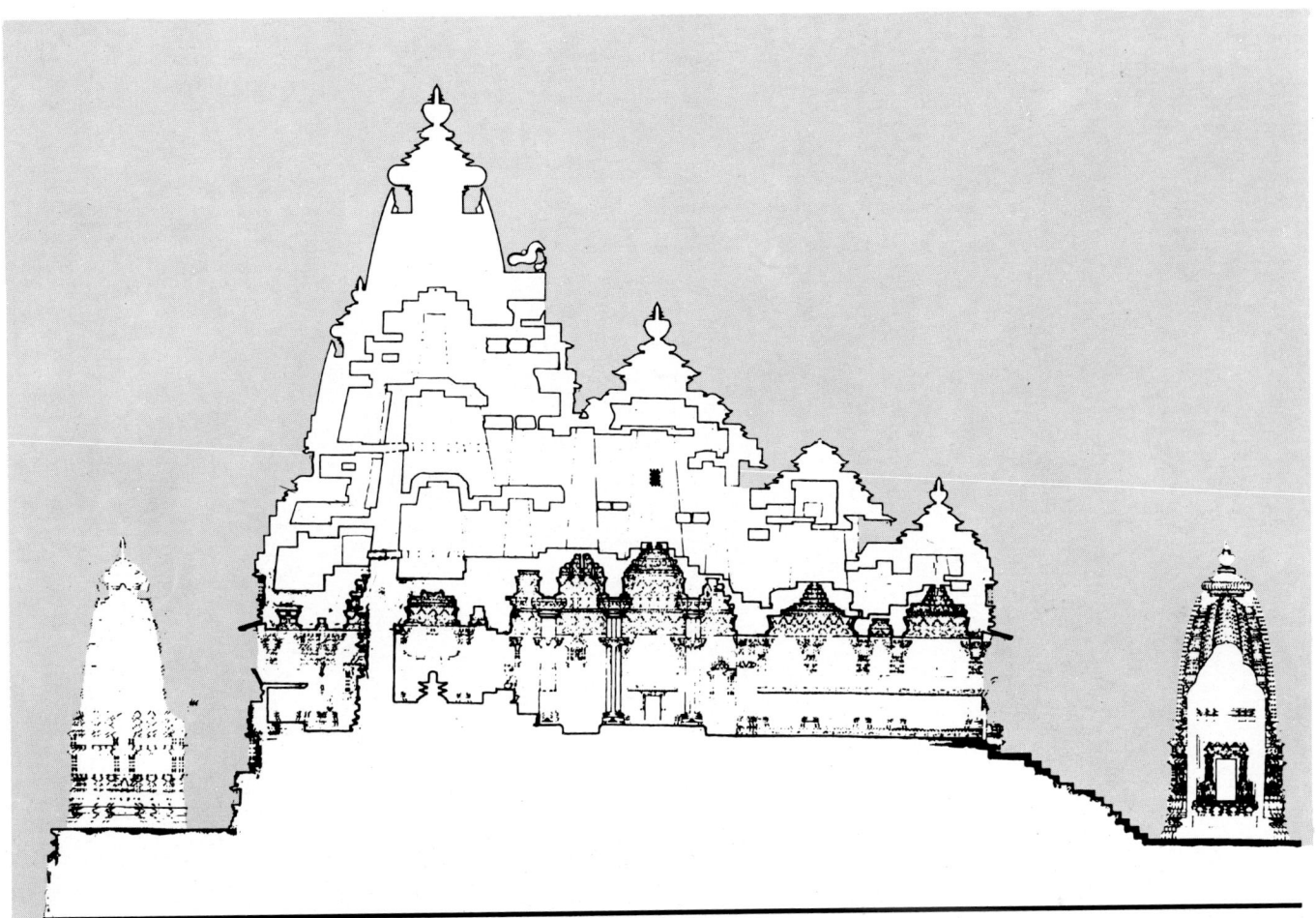

Fig 15.06 A sectional elevation of a Khajuraho temple, showing the similarity of the profile to a range of lofty mountains

The silhouette of a Khajuraho composition captures in all its magnificence the impression of a line of lofty mountain ranges (*Fig 15.06*). The functional part of the temple, the cella and its attached *mandapas*, are in contrast strikingly modest in volume. They are dimly lit spaces not unlike a cave carved out from the body of an artificially created mountain (*Fig 15.06*). Psychologically, the Indian builder was resolved to preserve the ancient traditional spirit of an Indian place of worship as a mysterious and dark cave, excavated out of living rock. In Khajuraho, he came as magnificently close to achieving this concept in reality as a man-made structure ever could.

Orrisa—A Study in Contrast

Returning once again to the contemporary development taking place in Orissa in the mid-eleventh century we find builders involved in the process of adding gargantuan structures to the now thriving pilgrimage centres of Bhubaneshwar, and the coastal holy town of Puri. Stylistically they clung to structures built over massive masonry walls with minimal openings. They were unable to emulate the Khajuraho example of airy open *mandapas*, with their huge pyramidical roofs seemingly delicately poised over a cluster of stone pillars. Of the more ambitious efforts of this period are the great Lingaraja temples at Bhubaneshwar (*Fig 15.07*) and the famous Jagannath temple at Puri, both of which are almost identical in style and construction.

◄ *Facing page*
*Fig 15.07 The great Lingaraja
at Bhubaneswar (AD 1000)*

The Great Lingaraja

The Lingaraja initially consisted only of a cella and a *mandapa*, each of truly gigantic proportions. The cella alone is 56 ft (17 m) square (*Fig 15.08*) and rises about 140 ft (43 m) into the sky. The *mandapa* is rectangular in plan, large enough to require additional support in the form of four massive central columns and its pyramidical roof reaches a height of almost 100 ft (30.4 m).

Unlike the temples of Khajuraho which were completely deserted after the fall of the Chandelas, those of Orissa, particularly the larger ones like the Lingaraja, developed into important centres of Hindu pilgrimage, and the elaborate ritual of Hindu worship is carried out in them to date. The Brahmin system of worship was becoming increasingly complex. It was no longer sufficient merely to obtain *darshan* or propitiate the gods by obsequeity or merely sing hymns in their praise. The concept of the temple as "house of God" was carried to its almost childish extreme. The Brahmins made it mandatory to feed, clothe, bathe and entertain the sculptured image of the deity, as if God himself was installed in the sanctum. The Brahmins now became the self-appointed trustees of this house, and the worshippers, the wilful servants of their idiosyncrasies.

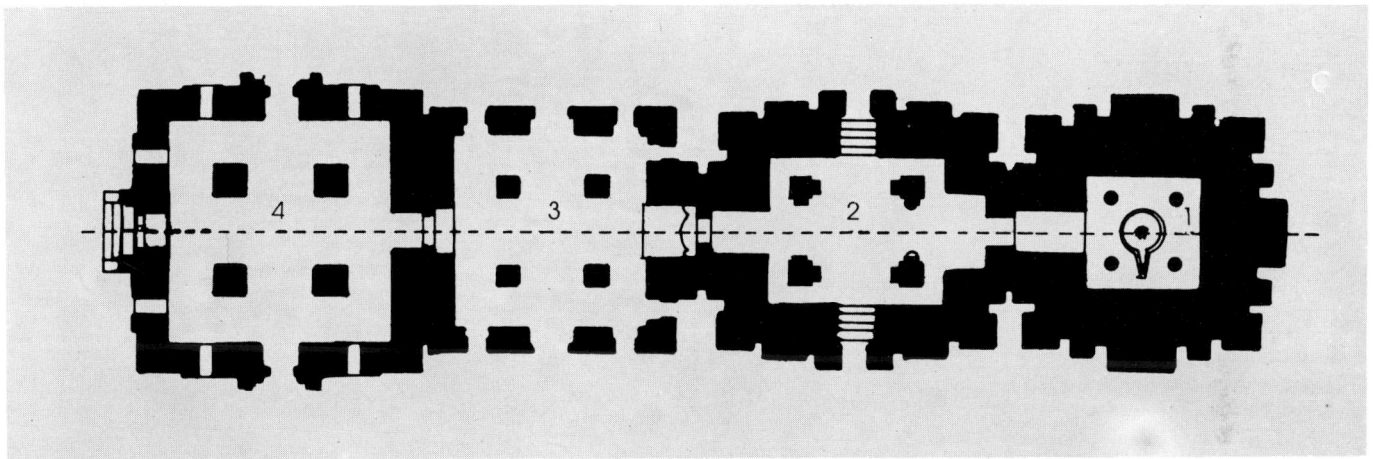

Fig 15.08 Plan of the great Lingaraja at Bhubaneswar. 1 Sri Mandir, 2 Jagmohan, 3 Nat Mandir, and 4 Bhog Mandir

In order that the various new rituals be carried out with due pomp and glory, two more halls had to be added to the existing *mandapa* or 'Jagmohan' (in Orissan terminology) of the Lingaraja. In the temples of Orissa these were called the Nat Mandir, where devadasis (literally, female servants of the gods) danced, ostensibly to entertain the gods, but more likely to soothe the carnal desire of the Brahmins. In the Bhog Mandir which was the next hall, food received from the houses of rich donors was supposedly blessed by the holy presence and a substantial part of it distributed to the poor. The image of God, enshrined in the inner sanctum, could look out upon the world only through one opening, that in the direction of the Jagmohan. For due fulfilment of the ritual of entertainment and blessing, the Nat Mandir and Bhog Mandir, could be attached only along the existing Deul Jagmohan axis. The square additional halls, not unlike the Jagmohan in plan, elevation and methods of construction added no doubt to the volumetric prominence of the Lingaraja temple but not to its architectural composition. The subsequent additions, being taller than the original Jagmohan, threw the entire composition out of balance (*Fig 15.09*). The silhouette, unlike the gradually descending sweep of the Khajuraho temple, became rather awkwardly convex since the Bhog and Nat Mandir soared higher into the sky than the Jagmohan planted in the middle.

Fig 15.09 A great temple, the Lingaraja at Bhubaneswar thrown out of balance by subsequent unimaginative additions

Looking at the myriads of other Orissa temples it would seem as if the craftsmen in spite of years of experience could visualize the temple only as a mere sum total of a series of unrelated units. The twelfth century Anant Vasudev temple is one of the rare examples of an Orrisan temple conceived and executed *in toto*. Though its silhouette has a clearly ascending profile, it is still as rigidly uncomprising as so many sergeant majors standing in drill (*Fig 15.10*) in sharp contrast to the fluid lines of a group of dancers that the temples of Khajuraho are. The temple, though smaller than its better known predecessors, the Lingaraj and Jagannath, stands on a more substantial plinth quite likely suggested by the massive stylobates of the shrines of Khajuraho.

➢ *Facing page*
Fig 15.10 The twelfth century Ananta Vasudeva temple at Bhubaneswar

The Raja Rani at Bhubaneshwar

It is obvious that a group of craftsmen from Khajuraho, or at least those well versed in its style had a hand in the construction of one of the more refined examples of Orissa. This is the so-called incomplete temple of Raja Rani. In plan the deul could be construed as a square placed diagonally in relationship to the Jagmohan with one of its corners rather than the broad face forming the link with the Jagmohan (*Fig 15.11*). In its elevational effect the Khajuraho device of attached *shikhara* quoins and *urusringas* is used to good effect. Though the temple spire acquired a pronounced sense of cambered grace and verticality, it still looks rather stiff and lacks the feline rhythm of its Khajuraho prototype (*Fig 15.12*). Hereafter, the only major remaining example of the Orissa craftsman's art is the great thirteenth century Sun temple of Konarak discussed later in its relationship to the great masterpieces of other regional styles.

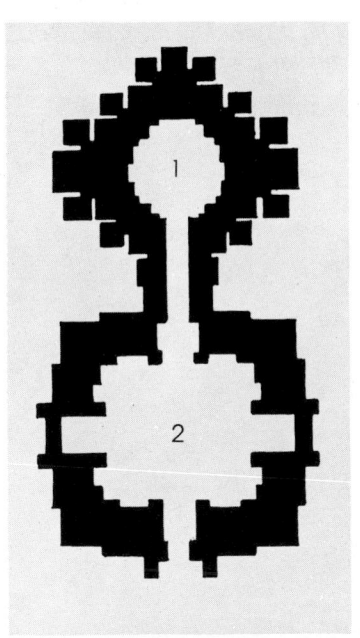

⋏ *Fig 15.11 The diagonally oriented plan of the Rajrani temple at Bhubaneswar. 1 Deul, and 2 Jagmohan*

Fig 15.12 Shikhara of the 10th century Orissa temple of Raj Rani inspired by the form of the Shikharas of Khajuraho

CHAPTER 16

The Hindu Jaina Axis of Gujarat

W̶e turn then to the western region of India, comprising parts of modern Gujarat, Kathiawar, Rajasthan and central India. This area was under the stable rule of the Solanki dynasty that had its capital at Anhilwad Pattan (near modern Ahmedabad). By dint of its location it controlled a long and jagged coastline abounding with natural harbours. Consequently, it was the focus of great trading activity between the eastern and western worlds, which led to the growth of a prosperous and flourishing mercantile community. The astute Brahmin was not to be left out of the general prosperity. Through the holding of "a great assembly on market day" he was able to persuade the rich community "to an agreement to set apart for the gods certain taxes. The shopkeepers were to give a spoonful of every kind of grain that was sold, while of cotton, each shop was to give as much as a man could hold in his hand" ... and so on.

Invasions and Earthquakes

With this abundance of funds and an artistic heritage going back to the time of the Guptas, the western countryside at one time was studded with fine examples of the building art. Possessed as these temples were of almost legendary wealth, they later became the target of the raids of Muslim adventurers. The most prominent of these raiders was Mahmud of Ghazni who attacked the area frequently in the early eleventh century. Among the many temples that he sacked the richest and most famous was that of Somnath at Veraval on the Kathiawar coast. Subsequently, in the early nineteenth century a devastating earthquake, with its epicentres somewhere in Kathiawar, reduced many religious edifices that were "splendours of sculptured stone" to "crumbling piles of broken and shapeless masonry." The few temples that have survived both the natural and man-made catastrophes are at Kiradu, Ghumli, Modhera and Sejakpur. It is evident even from these fragmentary remains that the Gujarat craftsman's peculiar genius lay in the building of sometimes exquisite, and sometime elaborate and grandiose *mandapas* for his temples.

The Temples of Kiradu

This propensity for designing the *mandapa* with exquisite care is noticeable even in the early temple of Somesvara (*Fig 16.01*) that stands today in a ruinous condition in the deserted village of Kiradu in modern Rajasthan. The tower over the *garbhagriha* is a rather ungainly pile of miniature *shikhara*-shaped turrets and triangular pedimented niches, erected over a richly though incohesively sculptured base. The *mandapa* is composed altogether more competently. Unhampered by an enclosing wall, it is an open structure held up on a series of columns (*Fig 16.02*). The arrangement of these is a departure from the traditional pattern of columns fixed along a rectangular or square grid. The heart of the *mandapa* is an octagonal

Fig 16.01 The pile up of minishikhara turrets that is the shikhara of the Someswara at Kiradu

space defined by eight tall and slender columns. Four aisles radiate out from the central octagon; three terminate in entry points to the *mandapa* and one leads to the *garbhagriha*. The plan is completed by setting the entire arrangement within a diagonally sited square, the four outer sides of which are further enriched by a series of rectangular offsets. The taller columns on the periphery of the central octagon are tied together by stone lintels and provide an intermediate ring of support for the pyramidical roof, constructed as ever by concentric rings of corbelled stone work. The zigzag outline of the *mandapa* creates not only a "vivid passage of light and shade" but also provides ideal oblong surfaces for embellishing with sculpture. The familiar inclined "asana" or stone bench that forms a heavy sloping band of stone over and all along the mass of the plinth reduces the appearance of the peripheral supports to mere dwarf columns.

The experience gained at Kiradu seems to have encouraged the builder to experiment with even more slender and elegant columns in the two Navalakha temples at Sejakpur (*Fig 16.03*) and Ghumli in modern Kathiawar state. The basic form of the *shikhara* and the overall plan in its diagonal configuration remains the same. However, lifted up as the temples are on a beautiful horizontally delineated cushion-like plinth (*Fig 16.03*) they acquire a grace all their own. This quality is further enhanced by the columns, some of which are circular in profile. With their pronounced horizontal bands of delicate tracery like sculpture, they seem as if turned out from an "embroidery stone lathe."

Fig 16.02 The remains of a once glorious open-pillared manadapa

Fig 16.03 A view of the front portico of the Navlakha temple at Sejakhpur

The Sas-Bahu in Gwalior

The architectural potential of the *mandapa* of the Gujarat style can today be seen fully realized in a pair of temples, standing cheek by jowl within the fort of Gwalior in Madhya Pradesh. These too, like the Teli-ka-Mandir nearby (described earlier) are today popularly known by the curious name of Sas-Bahu (literally, mother-in-law–daughter-in-law) temples (*Fig 16.04*). In the context of general Hindu structural timidity, only the Gujarat craftsman, with his skill in erecting columnar halls, could have conceived and built such daring structures. The volume of the Gujarat *mandapa* in either of the two temples explodes into a three-tiered structure consisting of a maze of open galleries and loggias, built around a central octagonal hall of over 80 ft height (*Fig 16.05*). To support the massive receding masonry of the pyramidical roof, the hall had to be provided with four central columns in addition to those of the octagonal ring. Though the dramatic quality of the interior space is somewhat diluted by the intrusion of these columns, it still makes a powerful impact. One cannot be certain whether the *garbhagriha* and *shikhara* attached to the *mandapa* were ever completed. Even if the builders had been able to erect one to match the proportions of the huge *mandapa*, no traces of the superstructure of either are extant.

Fig 16.04 Plan of Sas Bahu temple at Gwalior

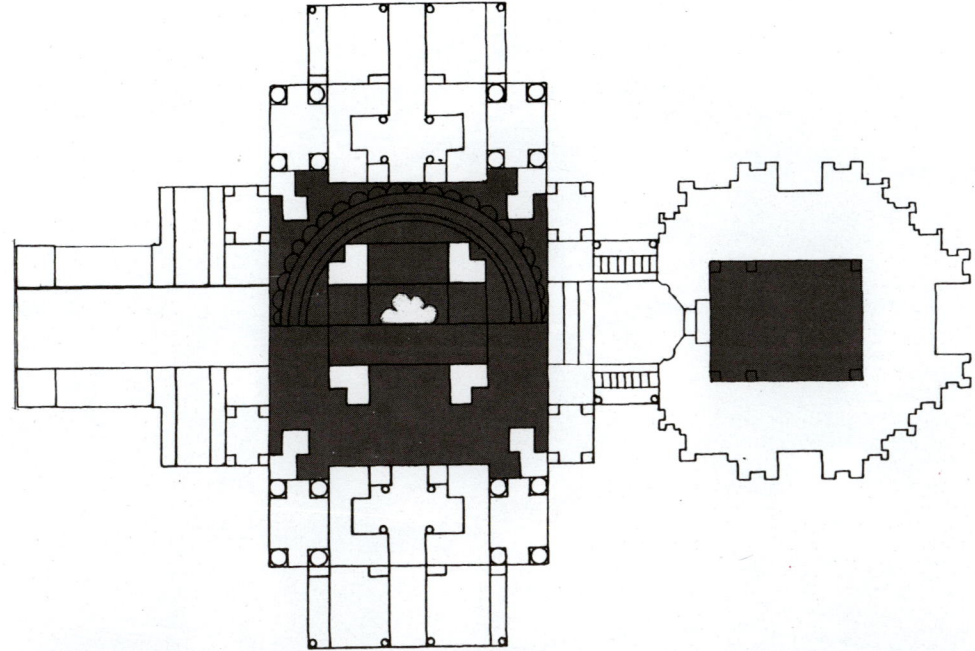

Fig 16.05 The mandapa of the Sas Bahu is an octagonal hall over 26 m high that was the glory of the Hindu builder's structural capabilities

The same fate has also overtaken a number of other temples in Gujarat where the *shikharas* have suffered great damage. Perhaps this could be reasonably attributed to the fact that a columnar construction for the *mandapa* akin to modern frame construction was, by its structural nature, better equipped to withstand the tremors of the great ninth century earthquake, than the solidly walled enclosures of the *garbhagriha* (Fig 16.06).

Fig 16.06 The larger Sas Bahu temple within the fort at Gwalior

Marble Glories of Dilwara

It would seem that in Gujarat and Rajasthan the tax set apart for the Jaina gods was greater than that set apart for the Hindu gods. This is not surprising. The affluence of the Jaina community was based entirely on commercial trade. Their rigid religious tenets of nonviolence prevented a follower of the religion even from being an agriculturist, leave aside a warrior, since in ploughing the fields he was sure to kill many live insects. The Jainas then had perforce gradually cornered commerce and trading as their main occupation. Ultimately, banking, trading and the power that went with it, was almost entirely controlled by the Jainas. Therefore, when they decided to build their temples at Dilwara near the modern hill resort of Mount Abu in Rajasthan, the expense involved in construction was of little consequence. While the Hindus had been content to build with the local golden brown sandstone, the Jainas were not to be satisfied with anything so modest. They chose the finest; pure white marble was to be their chief building material. The fact that it had to be hoisted from the valley three thousand feet below at considerable cost and labour, was no deterrent to the prosperous Jainas.

The Temple of Vimala Shah

The famous temple founded by Vimala, typical of the many more that were built and are now in ruins, is set inside a rectangular courtyard (*Fig 16.07a*). The enclosing cells are a series of niches adorned with images of *tirathankaras*, or the 'exalted ones' of the Jain religion. On the exterior the temple does not pretend to any architectural effects. In its plan, too, it is not dramatically different from the many Hindu temples of Gujarat. After passing through a domed entrance canopy and a hall filled with statuary portraying the Vimala family, one approaches the main body of the temple that consists of the familiar octagonally planned *mandapa* connected to the shrine at the other end by a transept having two parallel rows of pillars. The temple strives to achieve its most dramatic response from the sheer profusion of carving over the entire body inside and out. The most exquisite feature in this sumptuous feast of sculpture is the ceiling over the central octagonal ring of the *mandapa* (*Fig 16.07b*). The modestly proportioned dome constructed as ever by circular rings of oversailing masonry consists of eleven concentric friezes depicting patterns of figures and animals in procession. The lowest contains the forefronts and intertwined trunks of as many as 150 elephants. The dome culminates delicately in a pendant, suspended from the apex which, carved out of pristine white, almost translucent marble, looks like a fragile natural stalactite than a mere carving. Any real sense of architecture in the Vimala though is lost beneath the intricacy of carving and the profusion of detail (*Fig 16.07c*). So much so that even the natural delicate texture of marble is obscured having been fussed and fretted into excessive ornamentation. In spite of the fact that the ceiling of the dome really is like a "dream vision looming in the half-light" and the "figures

Fig 16.07 (a) The Vimala Shah Temple set inside a rectangular courtyard

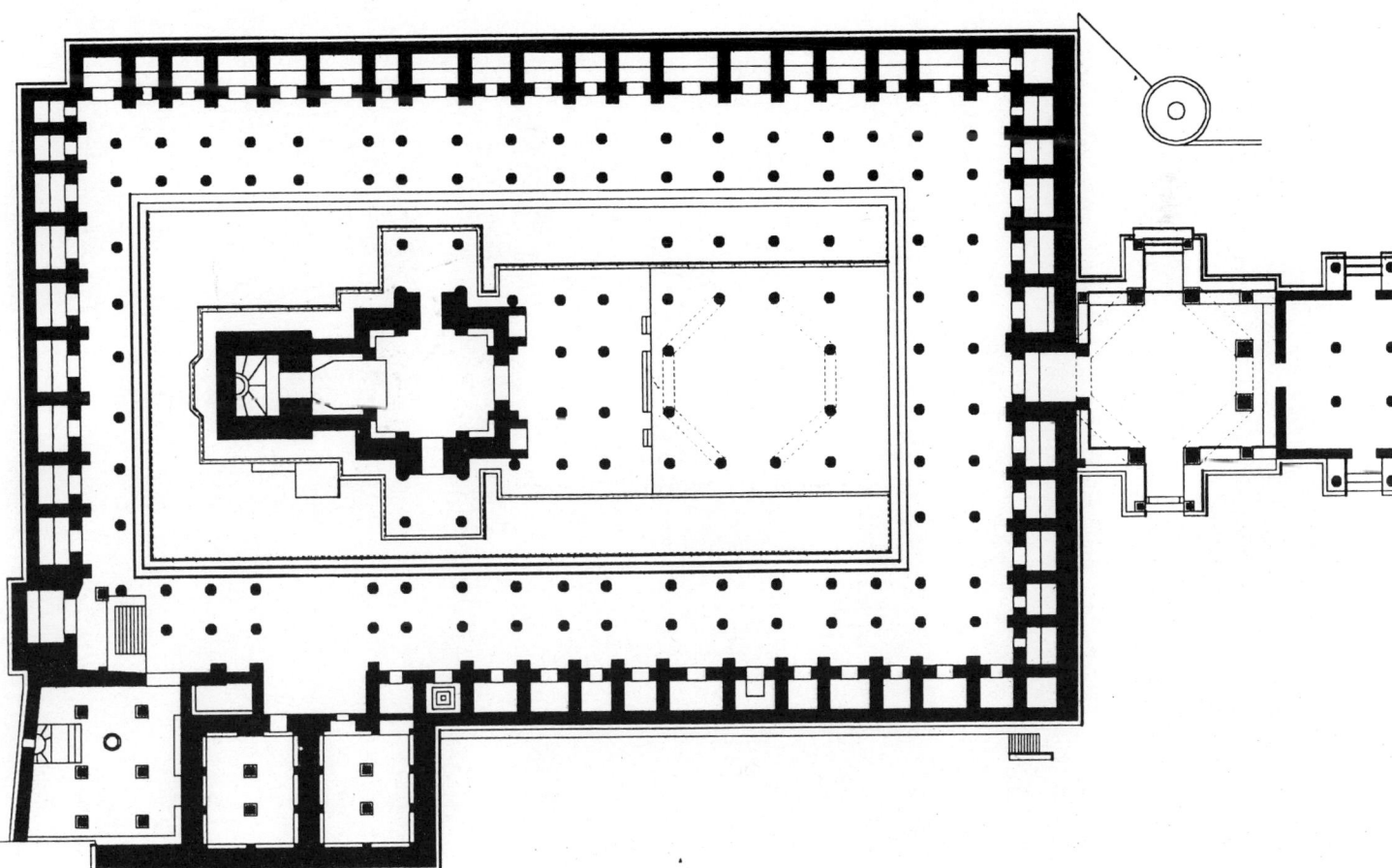

have the fragility of snowflakes," the perfection is of a rather mechanical character dependent as it is on the idea of creating impact through unceasing though perfect repetition. The overall impact of this shrine leaves one as cold as the marble of which it is created. Judging from the stern and forbidding visage of the god Mahavira (Fig. 16.08) installed in the cella, the designer had probably achieved the effect desired by the ascetic founder of the creed. However, at the same time there is no gainsaying the fact that "in its own ornate way the temple can be counted as amongst the architectural wonders (or rather curiosities) of the world."

Figs 16.07 (b) & (c) The exquisite mandapa and the ceiling of the Dilwara wherein a hard and adamant stone like marble is made to reproduce the fragility of snow flakes

Fig 16.07b

*Fig 16.08 The stern, and
forbidding Mahavira, the Jaina*

Fig 16.07c

Picturesque Jaina Cities

As may be judged from the description of the above temple, the Jaina religion which had developed into a sort of theological mean between Hinduism and Buddhism, never quite acquired a distinct architectural style of its own. Rather the Jaina designer would borrow the *shikhara* motif when it so suited his fancy and, with equal casualness, he would plant a mini-Buddhist stupa right next to it. Also, the mere antiquity of a temple, excepting of course obviously priceless edifices like that of the Vimala, did not add extra sacred value to the structure. In fact, the Jaina community seems to have frequently pulled down their old decaying buildings to replace them by newer ones, thereby leaving behind few examples of their earlier style. And so, what is largely left to us today of the Jaina tradition, is some very large and picturesque temple cities on the remote hills of Shatarunjya and Palitana in Kathiawar state (*Fig 16.09*). Most of the temples of these cities are of recent origin, and are an unsynthesised agglomeration of many styles of building including even the Islamic. It may, therefore, be pertinent to reflect a while on the theories of the Hindu builders that, unlike the Jainas, exercised sufficient control over the millions of temples built all over the subcontinent to impart to its vast array of architecture a distinct and discernible style.

Fig 16.09 A view of the Jaina temple city on the Shatrunjaya hill at Palitana in Kathiawar state

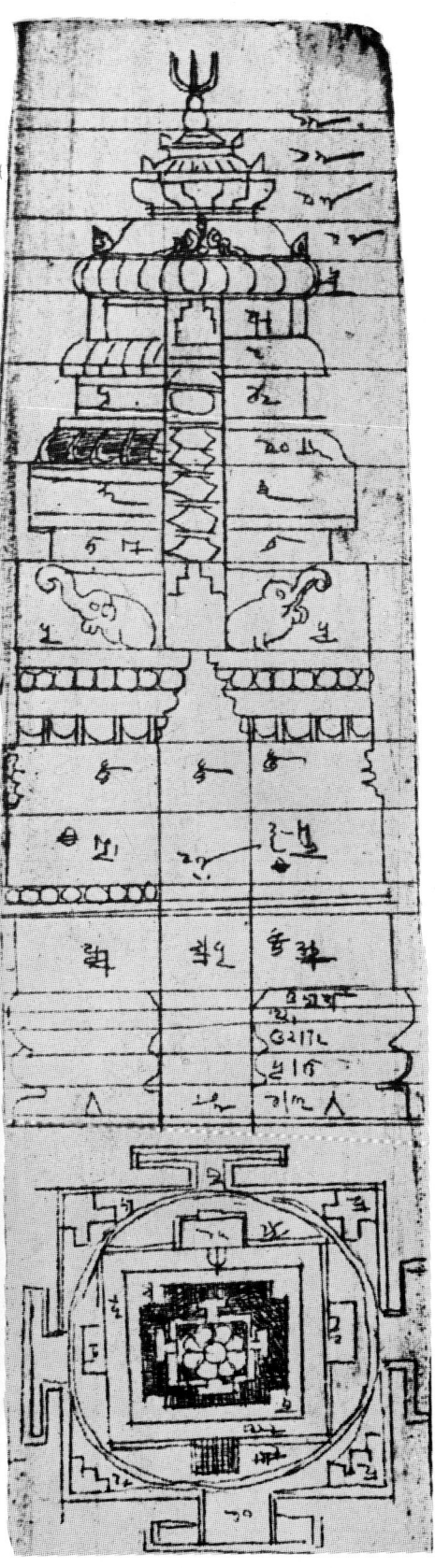

CHAPTER 17

The Mystique of the Vastupurushamandala

During the medieval ages in India there was no single dynastic power that could be considered the undisputed dispenser of cultural and artistic ideas. Yet, as we have seen, Indian temple design, other than what may at best be described as regional flourishes, displayed a remarkable unity of aesthetic purpose. True, the profile of the *shikhara* varied from region to region; temples were at some places raised on monumental plinths and at other hidden behind cloisters; the number of *mandapas* preceding the *deul* or *garbhagriha* would be anything between one to four; these would either be placed along a single linear axis, or in concentric rings; a northern temple would have a *shikhara*, and the southern one a *vimana*. Nevertheless, the evolution of each of these seemingly disparate elements was always motivated by a unified philosophy of design that permeated the remotest corners where the Hindu craftsmen were at work.

The Trabeate System

One obvious example of the Hindus' self-defined parameters is the universal adoption of the trabeate system of construction. It is strange that the structural economics of using the arch, vault, or other mechanical means which from the early Roman times were commonly employed all over the world, never became popular in India. The Indian builder, through centuries of development, was quite content to use the often wasteful system of vertical columns and horizontal lintels, of stone or timber, supported by brackets if necessary, to cover his roof or span openings. A fascile though poetic excuse of eschewing the use of the arch proffered by the Hindu builder was that "the arch never sleeps." As a rational result, however, the use of mortar was dispensed with, there being virtually no inclined pressures to distribute between the courses of masonry.

The Brahmins Make the Rules

This persistent uniformity of content and techniques is reminiscent too of the rather mechanical standardization of the ancient Indus Valley civilization. It could have been the result only of some kind of absolute and unbreakable rules governing the building craft. Not those laid down by a political power since, from the Gupta period up to the fourteenth century, when Moghul rule was established, there was no single political power of any consequence. A unified authority, if any, was wielded only by the priestly Brahmin class. They had, as we have seen, cleverly inveigled themselves into a position of controlling every facet of Hindu activity—bureaucratic, commercial, domestic and, of course, religious. No ceremonial activity, be it a birth, a death, marriage, reaching puberty, building a house or starting a commercial enterprise could take place without the consent of the learned Brahmins. This was generally granted after payment for the performance of elaborate rituals, due commissions in cash or gold and also, if the nature of the venture permitted, they received a continuing share of the profits. The art of building was not going to be an exception to the Brahmins' merciless iron web.

Cosmology and Architecture

The basic concepts of temple design were evolved through the creative activity of master craftsmen. It was nevertheless the Brahmins who began laying down elaborate and complex rules of layouts for building. These were intentionally made complex enough to be well beyond the comprehension of a layman or even a skilled craftsman and consequently beyond his rights to challenge. In the great Hindu compendiums of architectural rules—the *Vastushastra*—there is much that is deliberate esoteric verbiage. The very minutest of the acts of building, such as the layout, choice of site, testing of solid conditions and even the thickness of walls and columns were based not on technological, but rather on mythological and even astronomical consideration. This is often carried to a ludicrous extent. Formulae are drawn up to indicate how the thickness of a wall, or the span between the columns of a temple were to be determined from the day or time when construction commenced, the orientation of the temple, the caste of the major donor, and even the position of the stars in the sky. Needless to say, under such circumstances, the creativity of the building craftsmen had to be carried out within the theological disciplines framed by the priests in the *Vastushastras*.

The myriads of temples all over India are an eloquent testimony to the indefatigable Indian craftsman who was able to create such beautiful works of art, in spite of the countless restrictions placed on him. For this he must thank the integrity of the traditions of his craft which, being more ancient than that of the Brahmins, proved to be equally durable and tenacious. To him this sacred heritage was enshrined in the so called 'senis' or ancient guilds of craftsmen.

'Senis'—the Indian Craft Guilds

Even during the formative stages of the art in India, building craftsmen recognized not only the specialized quality of their skill, but also of its intrinsic value to society. As early as the seventh century BC when building activity was carried out with temporary materials like timber and unbaked brick, craftsmen had organized themselves into guilds to effectively guard their economic, technical and artistic interests. An association of this nature in its earlier stages would accept 'antevasikas' (literally, the boarders) as resident apprentices, to train them in the skills of their crafts. Over a period of time when the more laborious arts of cave and stone architecture became popular, the trainee required not only long periods of theoretical instruction, but also practical experience at sites often remote from urban centres.

Under such circumstances, heredity became a custom. The skills of the trade were conveniently passed on from father to son. At the earliest opportunity the son was given a chisel and hammer and trained to give shape to a block of rough quarried stone, carving of finer details being left to the father. Often during the construction of a single enormous venture, the son would acquire the skill of his father, to take his own rightful place as a master craftsman. The builder, though working for the property owning wealthy class, himself rarely acquired any immovable property that he could leave to his sons. The convention of heredity, apart from its practical function, assured the skilled artisan that his son would at least continue to retain his status in society. The transfer of skills to his offspring was the craftsman's insurance policy on his skills.

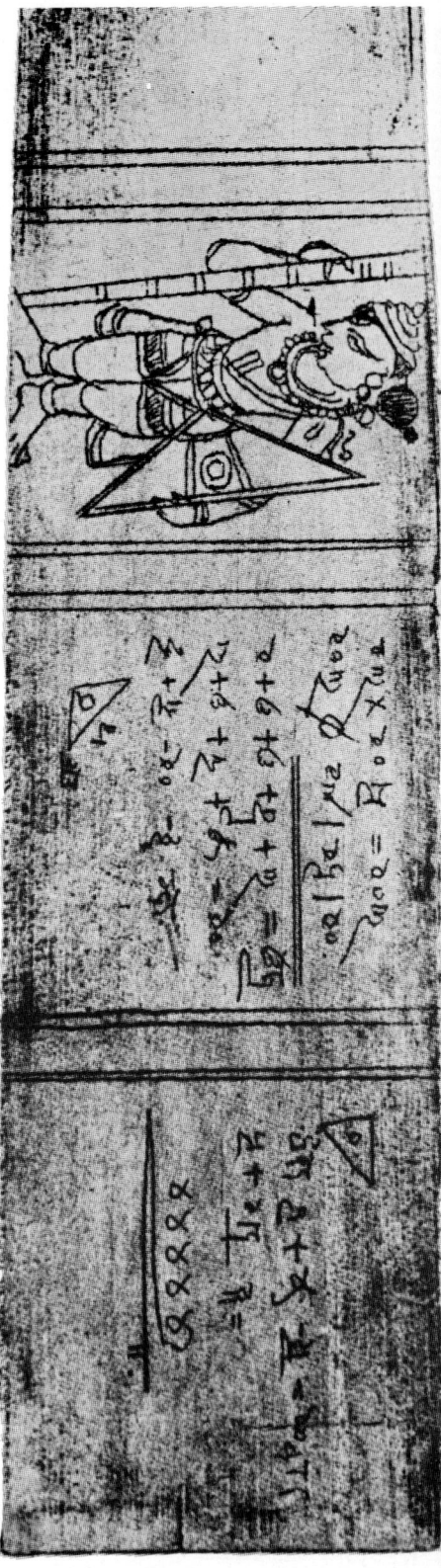

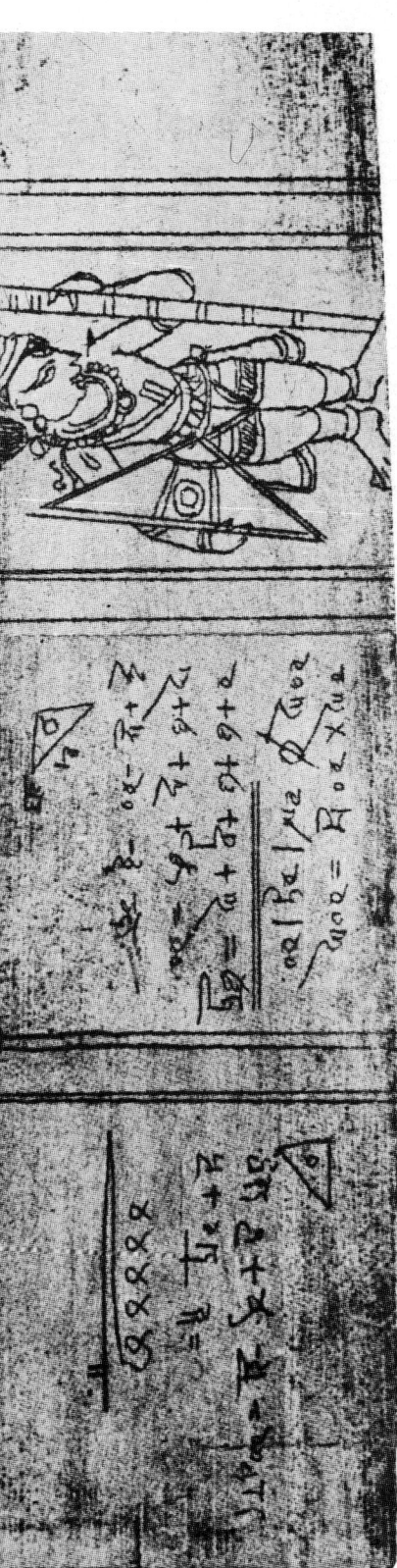

Schools of Style

On some occasions the magnitude of their task would force a group of craftsmen to settle down near the site of their labours for a generation or more. It was such a settlement at a single site, rather than purely academic thought, that resulted in the growth of an art centre, giving rise to a school and a style of architecture. So it was, that a group of craftsmen working continuously for over a century on the building ventures of the Chandelas, gradually evolved the characteristic Khajuraho style. Or those concentrated around Bhubaneshwar formulated the Orissan mode of building.

In spite of their rigorous and obscure rules, the Indian cannons, or rather ordinances of art (which had to be memorized by one ambitious enough to become a master craftsman) had their own positive value. Architect Le Corbusier claims that his theory of proportions "the Module, made the bad impossible and the good easy," has rarely been put to test. The practising contemporary architect never quite took to it as his design gospel. Percy Brown's assessment of the Indian counterpart 'the Mandala,' that "if faithfully followed, would make failure impossible," is, however, not an empty claim. It is a fact borne out by the quality of Indian temple activity through centuries, and spread over territories separated by thousands of miles.

The Magic of the Square

Confronted with the myriads of exotic forms of the Hindu temple it is difficult to believe that the crux of the guiding philosophy of design of the Mandala was the square; most basic, rational and elementary of all geometric forms. For the earthly abode of God there could be nothing but the most perfect of forms, obviously either the circle (a polygon of infinite sides), or a square (a polygon of equal sides joined at right angles). The Hindu chose the square. The circle, which had been adopted by the Buddhists for reasons of their own, to the Hindu represented movement, as inherent in a wheel, or in the ever-mobile spherical forms of the universal cosmos. The Hindu's aims were quite contrary. His temples were meant to be permanent abodes for his otherwise heavenly and elusive gods. Their images had to be installed in shapes symbolizing stability rather than mobility. The square fulfilled his aims far more appropriately.

The square Mandala (at best "divine chart" in English) was divided into so many equal squares—that containing 64 or 81 (*Fig 17.01*) being the most popular.

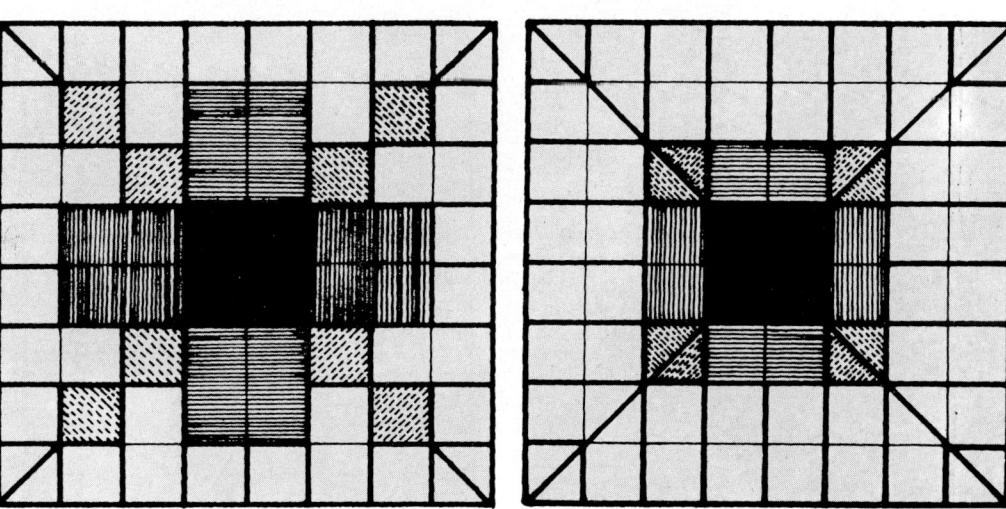

Fig 17.01 The square Mandala or divine chart of the Hindu builder

The priest then invested each of the squares with metaphysical and magic powers by locating an individual deity in each. The position of each subdivided square in the total represented the power or otherwise the deity attributed to it. Thus, Brahma, the supreme god, creator, preserver and destroyer, invariably occupied the central square or group of squares. Lesser deities were posted in the four corners (the germ of the *pancharatna* plan) and more minor ones filled up the balance (*Fig 17.02*). To invest the square with a human quality, apart from its divine one, it was shown as being able to accommodate within itself a human figure, though in a contorted yogic pose (*Fig 17.03*).

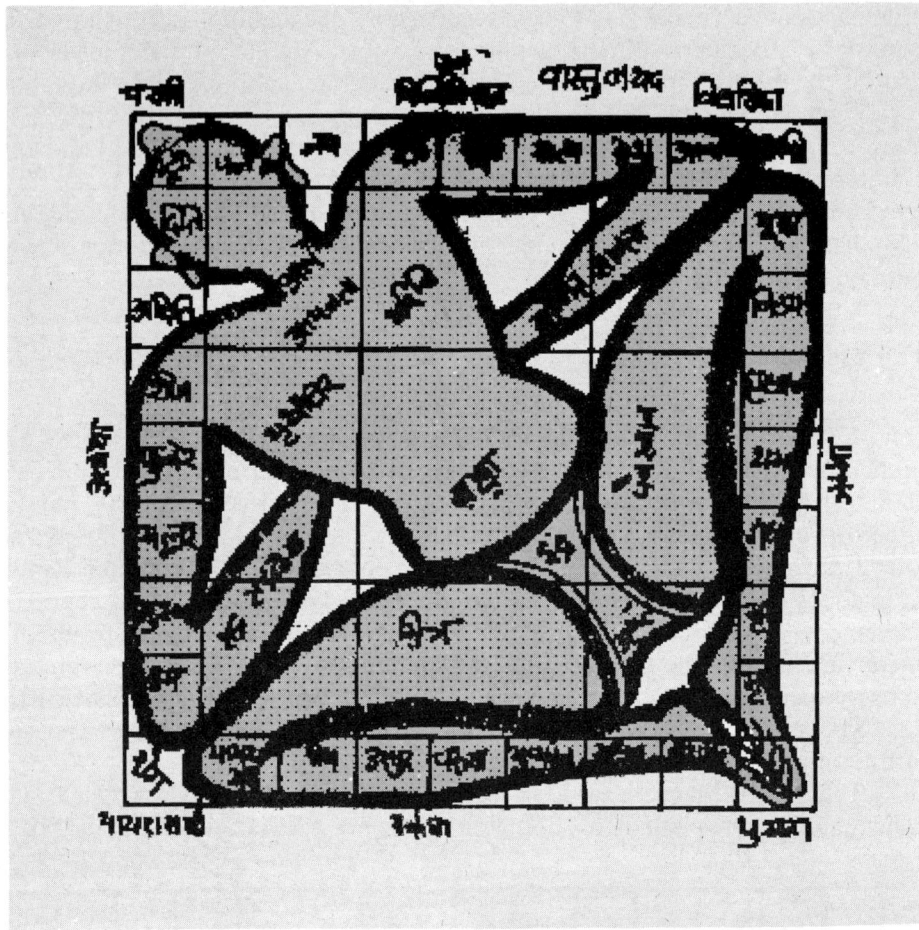

Fig 17.03 The Hindu Vastu-purshamandala consists of a human figure in a Yogic posture, enclosed within a square

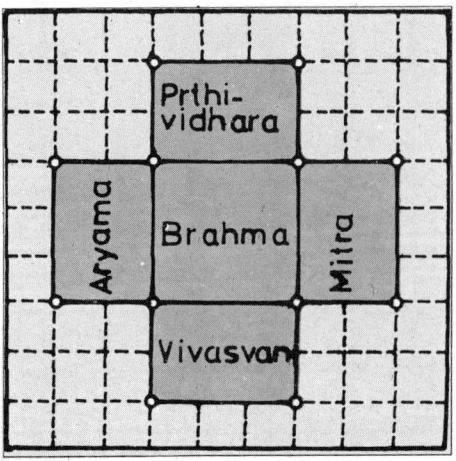

Fig 17.02 The various deities around Brahma in the Mandala

Humanism in the Square

In investing the dimensions of architecture with this human quality, the Hindu mind had preceded Leonardo da Vinci who, hundreds of years later, demonstrated how a circle could be drawn to contain a human figure with outstretched hands and feet (*Fig 17.04*) or even Corbusier's theory of the Modular that was built around human dimensions (*Fig 17.05*). The Hindu priest's humane touch to architecture was, however, quite different. In this case it is clear that it is not that the square was really inspired by the human body, rather the form of the *purusha* (human) was made to suit the abstract idea of the square as the geometric form supreme.

Fig 17.04 The renaissance concept of the human figure in a circle

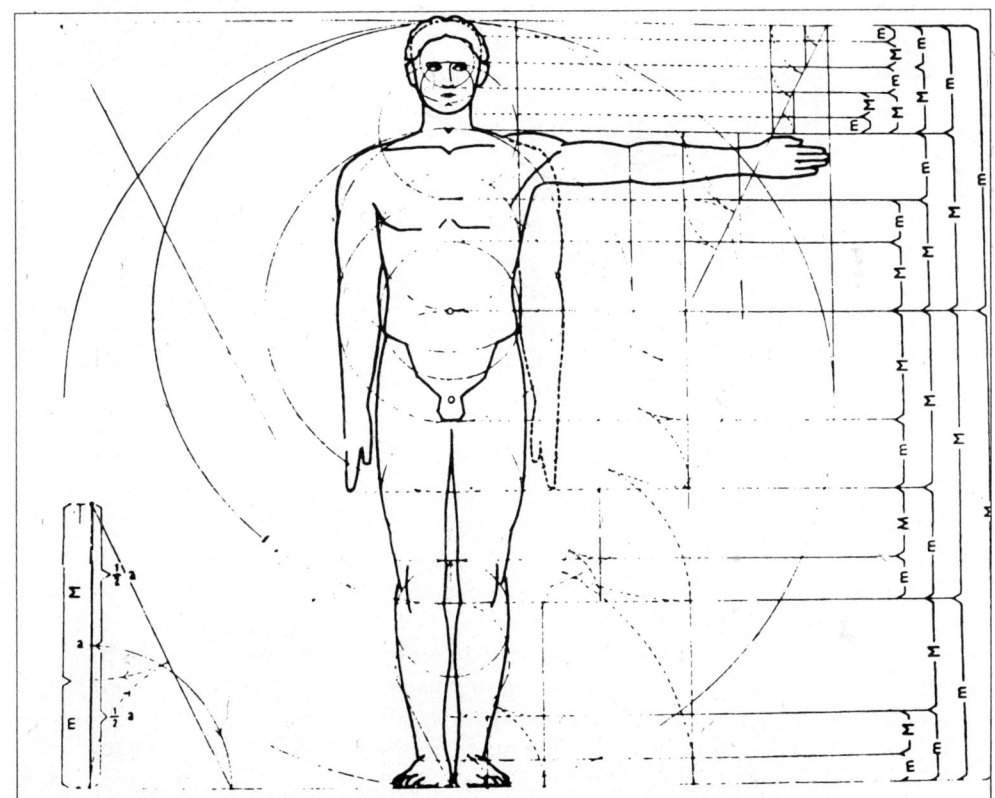

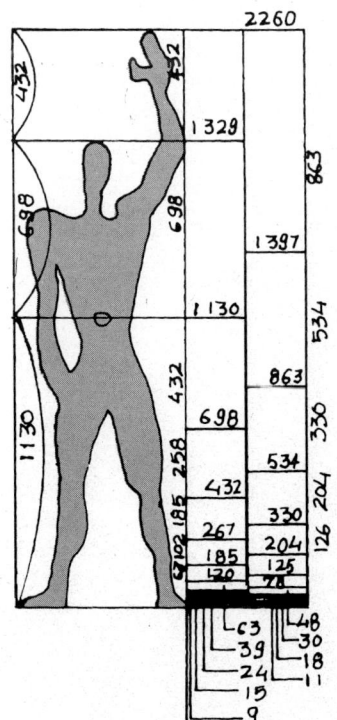

Fig 17.05 Corbusier's modular and the human scale

Having acquired magical, geomantic as well as human properties, the emerging chart called the Vastupurushamandala (literally, the magic chart of the architecture of the supreme man), was now fit to be transformed into an architectural ground plan for a temple. In its simplest form the outer enclosure of square could denote the thickness of the walls of the *garbhagriha* (*Fig 17.06a*). On another scale, four central squares could constitute the inner cell surrounded by an enclosure of 12 squares which became the walls and the next 16 to 28 the *pradakshinapath* and outer wall, and so on (*Fig 17.06b*).

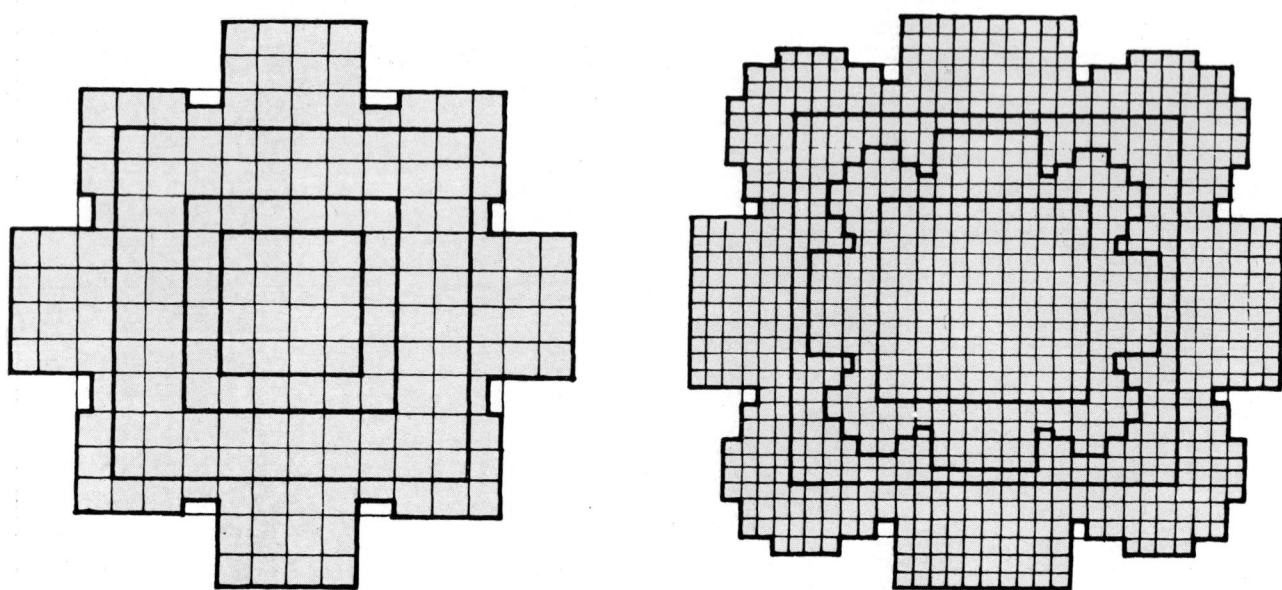

The elementary division was only the first step towards the building up of a rather complex system. With a bit of artistic license, the Mandala could be expanded to generate the most elaborate of forms, the basic unit of these always being the square (*Fig 17.07*). And so the large square was subdivided into thousands of small squares by the architect. This enabled him to add sculptural richness to his wall surfaces by the simple expedient of treating the Mandala virtually like a graph paper, adding a unit here and setting one back there, not unlike the modern architect's structural grid. It was with an artistic manipulation of the grid ordained by the Vastupurushamandala that the Indian architect created the profusion of exotic shapes and forms that are the temples of India.

But, again, though the very form of the *shikhara* defied measurement through right angles and squares, the theoretician left little allowance for blunders in its profile. He ensured that the curve of this essentially bamboo shape was translated accurately into stone, by meticulously plotting it out against the curves of more precise geometric forms like the ellipse (*Fig 17.08*). Furthermore, the entire structure was classified into many smaller elements, each with its own defined function, and with names, analogous to parts of the human body.

Figs 17.06 (a) The Hindu Vastupurshamandala (b) Transformed into architectural plans

'Alive' and 'Dead' Materials

To make such an abstract of geometry work in building practice, the Hindu cannons propounded a truly unique theory of building materials and their structural properties. Materials were classified either as 'alive' or 'dead'. The wood in a growing tree was 'alive,' and its structural properties were therefore to be respected.

Fig 17.07 Plan of the Ambarnath temple.
1 Garbhagriha and 2 Mandapa
➤

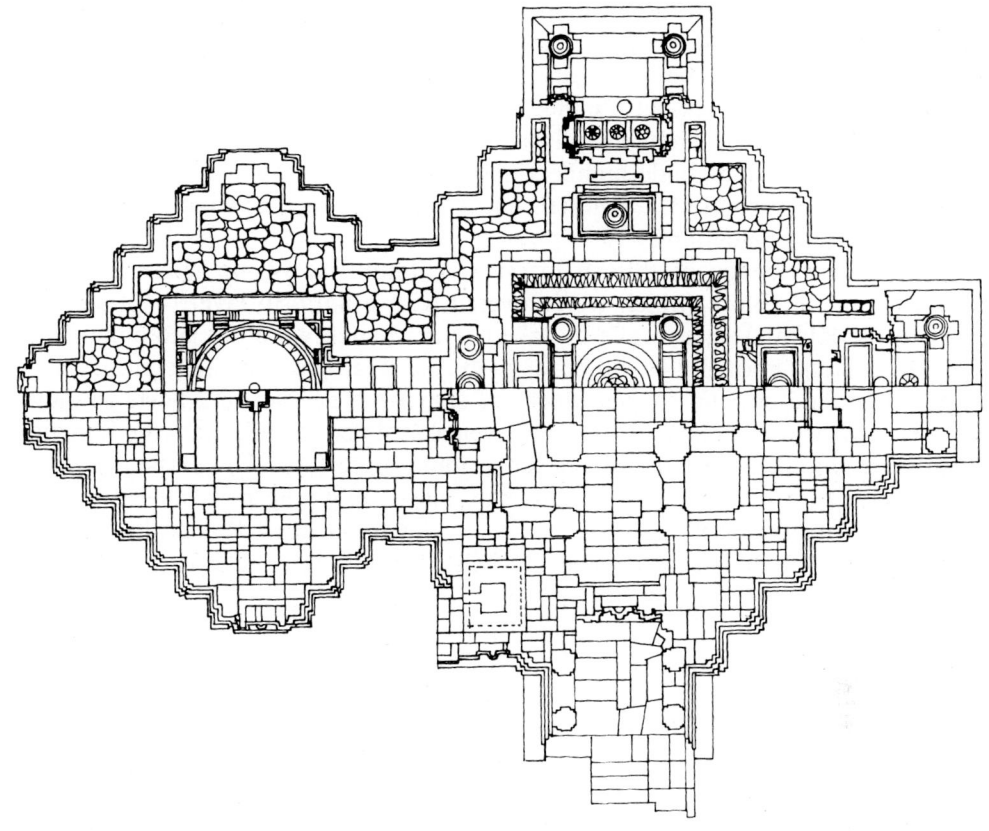

Fig 17.08 The mathematics of the parabolic shikhara

1 Kalasa
2 Amla
3 Chhapra
3a Bhumi
3b Amla
4 Bada
4a Jangha
4b Barandi
5 Pista
6 Rahapaga
7 Paraghar
8 Ghanta Kalasa
9 Pida

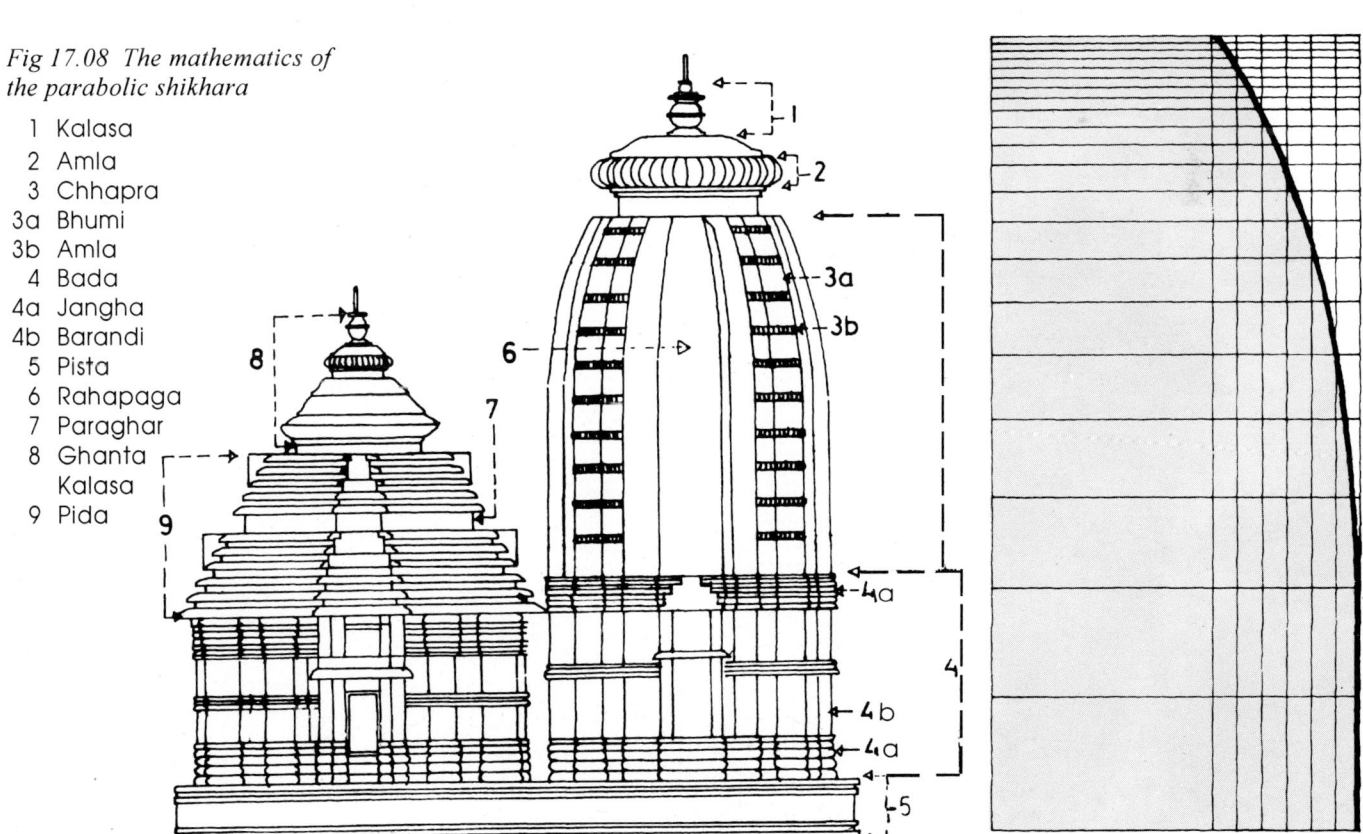

Once felled, the timber obtained from it, however, was 'dead'. It became a material to be used and shaped according to the purely artistic inclinations and whims of the architect and the technical capacity of the craftsman. No wonder then that the more ductile form of rock-cut architecture proved to be the true metier for the expression of the Indian architectural genius. On the other hand, in dealing with structural buildings, the Hindu architect never worried much about exploiting the 'true' structural potential of a material, but was more concerned with using it to fulfil his aesthetic intentions.

We have evidence of this in massive sculptural works in stone like the Kailasa temple at Ellora, which were covered with a uniform layer of plaster to look like the Himalayas; or walls of baked bricks which were cut and embossed as blocks of stone normally would be, as in the brick temple at Bhitargaon. Earlier stone masons had cut and shaped blocks of stone into needles and posts as they would be sawn out of timber. These were even assembled together into the famous Buddhist railings, with a mortice and tenon joint, more logical to carpentry than stone masonry. The patient Hindu craftsman, intent on giving permanent form to his visions, was not unduly bothered by such technical niceties. Rather, he created a peculiar kind of drama by forcing the will of his abstract artistic theories on materials not readily suited for them.

CHAPTER 18

The Splendours of Medieval India

The Indian craftsman after centuries of familiarity and practice, from the tenth century onwards took up the challenge of erecting great temple cathedrals in different parts of north India. Three of the more significant achievements in the west, central and eastern regions of India range from Modhera in Gujarat to Khajuraho in Madhya Pradesh and Konarak in Orissa. Being representative of the architectural qualities inspired by the regional styles they are worthy of comparison, as being the finest achievements of these three great schools.

The Sun Temple, Modhera

The acme of the Gujarat craftsman's art was the temple of the Sun God of Modhera (*Fig 18.01*) erected in 1026, on a site 18 miles south-east of modern Patan. Temples to the Sun God, even in distant parts of India, have had a curiously uniform pattern of development and decay. The temples of Martand, Konarak, Orissa and Suraj Kund in Delhi all seem to have been constructed on specially selected sites away from the humdrum of daily urban life; they are unusually ambitious efforts, and each one has lain deserted and ignored by devout pilgrims for the past many centuries. Nevertheless, even in their derelict splendour, they bear testimony, if nothing else, to the ambitions of both the builder and his feudal lords (*Fig 18.02*).

Built most probably under the patronage of the Solankis, the Sun Temple of Modhera, too, is no exception to the ravages of time. "What, at one time, may have been a site endowed with considerable sacred sanctity, is today a village of mud huts," close to the ruins of a once grand Sun Temple. The various constituents of this temple, the huge bathing tank, columnar *sabha mandapa*, a peristyler

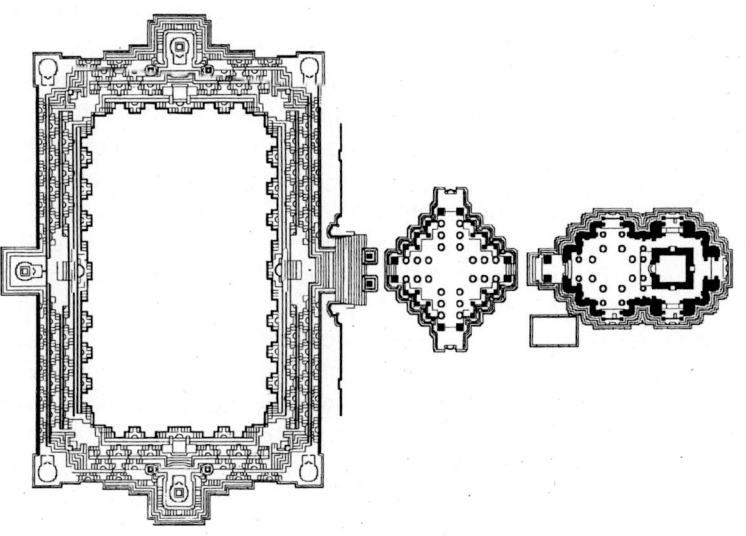

Fig 18.01 Plan of the Sun temple at Modhera in Gujarat. 1 Prasada, 2 Gudha mandapa and 3 Sabha mandapa

assembly hall and shrine, are knit together into an "entirely organic plan in the relation of all the parts of the shrine to the whole, and its functional arrangements of all the architectural accessories of religious worship."

The huge rectangular tank of ritualistic ablutions lies beneath the main eastern approach to the temple. Water was contained within buttressed walls that are a maze of steps and terraces, interspersed with mini shrines—an architectural and engineering masterpiece in its own right. A broad flight of steps takes one up to what was a *torana* archway resting on two gracefully moulded columns. The profile of the capital shaft and base of each of these is all contained within one sweeping rhythmic curve (*Fig 18.03*) typical of the gracefully sensuous style of Modhera. Immediately behind the *torana* lies the *sabha mandapa* (*Fig 18.03*), planned in the usual Gujarat style around an octagonal pillared space. Sloping seats and dwarf columns are set in rectangular offsets around its periphery. As in the other temples of Gujarat, it is this *mandapa* that is both the functional and visually vibrant heart of the whole complex. The shafts of each of its columns are exquisitely embroidered with scrolls of ornamentation interspersed with sculptures of gods and goddesses. The pillars on the periphery of the central octagonal space are laced together by delicately carved interlinked sinuous brackets aptly referred to as 'flying' *toranas*. The rest of the temple comprising the *garbhagriha* and attached *mandapa* has virtually vanished. It was a conventional walled structure. Its ornamentation, though, was in perfect consonance with the graceful rhythmic lines of the main *mandapa*. This entire edifice, raised as it was over a massive rectangular stylobate, must have been a fitting tribute to the great god that it purported to worship and honour. "To see even the ruins of this noble monument not only rising like an exaltation but mirrored in the still waters below, is to feel that the creator was more than a great artist, but a weaver of dreams."

Fig 18.02 A view of the existing mandapa of the Sun temple at Modhera

➤ *Facing page*
Fig 18.03 The sabha mandapa and the sensuous pillars of the Modhera

The Sun Temple, Konarak

In paying their homage to the Sun God, the dreams of the master craftsmen of Orissa, were of a different nature altogether. They decided to house their Sun God literally as Indian mythology has conceived it: "blazing his way through the heavens on the chariot of time pulled by seven leaping and prancing magnificent horses." This remote site was located on the sea coast of Orissa, away from the traditional and thriving religious centres of Bhubaneshwar and Puri.

To simulate the appearance of a *ratha*, or wheeled car, the longer sides of the terrace over which the temple was erected were ornamented with reliefs of twelve meticulously carved massive wheels, more than 10 ft (3 m) in diameter. Each of these giant wheels is a faithful reproduction in stone of its original, complete with a hub, timber pins and spokes, all profusely and meticulously sculptured (*Fig 18.04*). The parapets on either sides of the flight of steps rising up to the entrance are in fact a row of life-size sculptures of "richly caprisoned steeds, rearing and straining in their harness, as they strive to drag the great bulk (of the temple) along." Over the main platform are placed the combination of the *Jagmohan* and *Deul* temples. On another high plinth facing the majestic flight of steps leading to the shrine is situated the Nat Mandir which is a modest sized replica of the *Jagmohan*. The entire complex (*Fig 18.05*), including also a subsidiary shrine to Ramachandra and a refectory of utilitarian design, is set within a vast enclosure of 865 ft (263.6 m) by 540 ft (164.5 m). Presumably, the enclosing wall, the superstructure of which has altogether vanished, was provided with three pyramidical roofed gateways located in the centres of its eastern, northern and southern walls.

Fig 18.05 A conjectural restoration of the grand complex of Konarak as it must have been in its full glory in AD 1250

◁ *Facing page*
Fig 18.04 The stone wheels of the Sun God's chariot each of which is more than 3 metre in diameter and profusely sculptured

Failure of Structure

The *shikhara* over the *deul* containing an image of the Sun God carved in chlorite was envisaged to ascend to a height of over 200 ft (60.8 m). The substantial masonry necessary for such a grand *shikhara*, was wide enough to enclose three subsidiary shrines, each approached by an individual external staircase. Even so, the ambitions of its chief patron, King Narsimhdeva (AD 1238–64) were well beyond the technical skill of his craftsmen. The massive *shikhara* is today only a "battered and broken pile of masonry." The tower most likely never rose to its intended height, since the foundations started sinking before its completion. The 100 ft (30 m) high cubic mass of the *jagmohan*, built over massive and thick stone walls was given additional support in the form of four massive central piers, spanned by stone lintels reinforced even with wrought iron beams. In spite of these precautions, sometime in the nineteenth century, the interior of the hall was made inaccessible as it had to be completely blocked with rubble from inside to prevent its corbelled roof from caving in.

Triumph of Sculpture

Even from the ruins of the great Sun Temple, it is obvious that the massive proportions of the temple were no deterrent to the sculptor. In addition to the traditional massive reliefs that cover the entire outer surface, he also proceeded to adorn the gigantic pyramidical roof of the *jagmohan*. As a result of its massive size, the diminishing tiers of its roof are interspersed with what become broad platforms rather than mere steps. Inspired by the overpowering scale of the architectural backdrop, the Indian sculptor planted full-bodied statues of heroic dimensions along the edges of the lowest platform (*Fig 18.06*). Each of these represents a musician gracefully contributing his mite to a silent but mighty orchestra in stone. In spite of the fact that this was the Hindu sculptor's first attempt at embellishing an architectural facade with completely free standing statuary he executed the job with his usual consummate mastery. It is clear then, that the remains of the grand temple complex as they stand today are more a tribute to the Indian craftsman's sculptural skills than his structural ingenuity.

Fig 18.06 (b) A view of the existing ruins of Konarak

Fig 18.06 (a) One of the gigantic larger than life, full-bodied sculptures on the eaves of the Jagmohan of the Sun temple at Konarak.

The Kandarya Mahadev, Khajuraho

In contrast, the builders of Khajuraho, both artistically and structurally, were more successful in erecting their *piece de resistance*—the temple cathedral of Kandarya Mahedev. In expanding the plan of the temple to contain four, instead of the more usual two compartments, the designer never lost sight of the necessary interrelationship of each to the whole (*Fig 18.07*). The result—a perfectly balanced and composed elevation, its components like so many masses gracefully interlinked with each other (*Fig 18.08*). All the elements of a full bodied temple containing the *garbhagriha*, a surrounding *pradakshinapath*, *antaralaya*, *mandapa* and *ardha-mandapa*, become the cells of a unified organism. The first four are compactly enclosed within a continuous peripheral wall pierced only by beautifully crafted balconies. Just enough light is admitted through these to bring the profusely sculptured surfaces of the interior to life (*Fig 18.09*). The roof of the *ardha-mandapa* on the other hand is held up on columns, and is an appropriately extended version of the entrance portico.

Fig 18.07 Plan of the great 10th century Kandarya Mahadeo temple, Khajuraho. 1 Garbhagriha, 2 Antrayla, 3 Mandapa and 4 Ardha mandapa

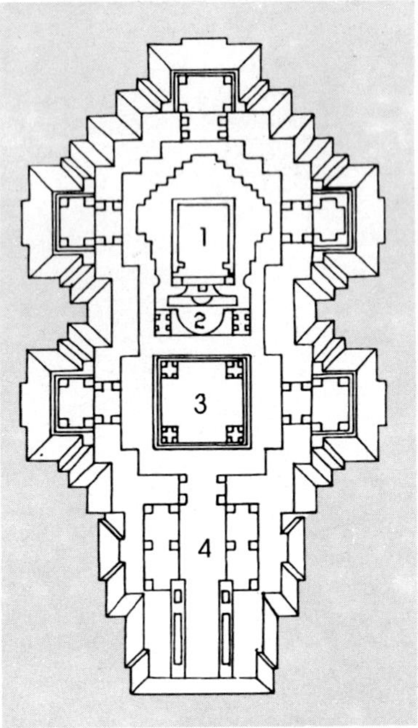

Fig 18.08

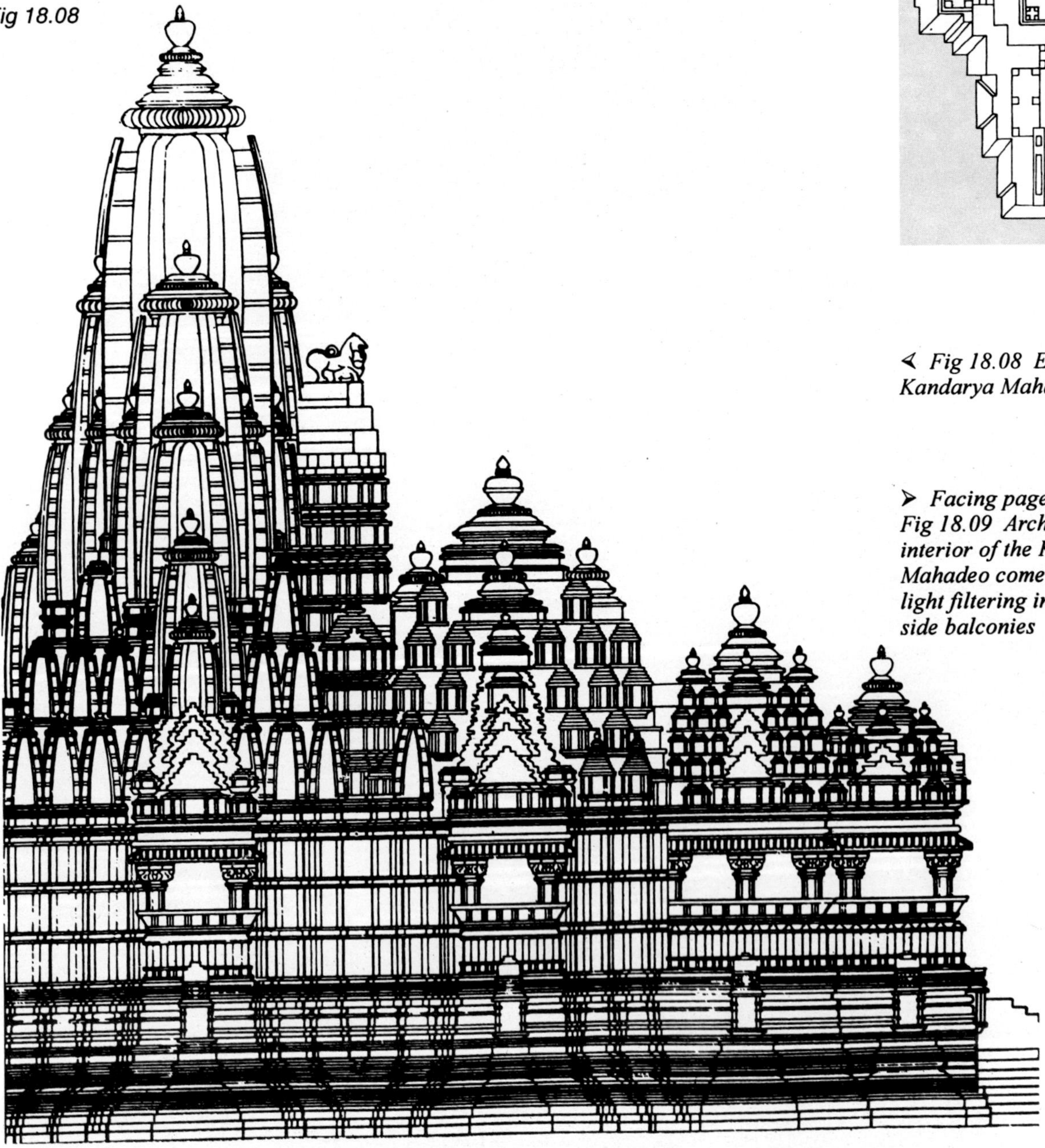

◄ Fig 18.08 Elevation of the Kandarya Mahadeo, Khajuraho

➤ Facing page
Fig 18.09 Architectonics of the interior of the Kandarya Mahadeo come to life under the light filtering in through the side balconies

It is true that the height of the temple spire (116 ft) (35 m) is little more than half that envisaged for the great Sun Temple at Konarak. But the quadruplicated application of the *urusringas* described earlier, and the soaring vertical lines of the *shikhara*, accelerate the vision along the crescendo of curves, reaching out to the heavenly gods, as it were. These all create an illusion of height that Konarak, even if completed, may well never have achieved. The vertical volumes of the *shikharas* of Khajuraho take off from an adequately cushioned base of marked horizontality, which is preserved right up to the level of the balconies in the outer walls. The dark voids of interspersed balconies and the open pillared *mandapa* in front create the impression of a massive pile of masonry, poised delicately over slender columns, adding a subtle touch of lightness to the virtual mountains of stone (*Fig 18.10*).

Fig 18.10 A massive pile of masonry delicately poised over slender columns; the temple of Kandarya Mahadeo

The Erotica of Hindu Temples

Apart from all their points of difference, a feature that is common both to Konarak and Khajuraho, is the remarkable profusion of the now famous erotic sculpture of the temple walls. To a certain degree, erotic or at least highly suggestive sculpture and painting are common in Indian temples, almost from the earliest times. They can be passed off as expressions of the Indian craftsman's "love for life in all its aspects." But the erotica of Khajuraho and Konarak is another thing altogether. The elaborate and extremely graphic presentation here is not restricted merely to amorous embraces or copulation between man and woman, but includes also, what progressive terminology would describe as free group sex, and the conservative as downright perverse and degenerate orgies.

The fantastic variety and exuberance of such erotica has prompted comments covering a wide spectrum of thought. To the devout defenders of the moral purity of Hindu faith "this evil erotica is purposefully perpetrated ugliness to ward off the evil eye" or "a device to highlight the contrast between the evil outside and purity of the inner sanctum"; the latter supported by the fact that erotic sculpture is to be found only on the outer walls and never inside. Some believe that the erotica was inspired by various tribal tantric rights which, along with their unlimited sexual license, had been absorbed into Hindu mythology. To the more outspoken, however, it was nothing more than the downright immorality that had crept into the courtly life of the kings and the religious life of the priests. This section of opinion derives peculiar satisfaction from the conjecture that the once powerful Chandelas vanished from the scene so quickly after building such great temples, only because of their excessive indulgence in such immoral acts.

A Freudian Catharsis

On the other end of the scale are the Freudian thinkers. The theme of sex and other libidinous excesses that are so vividly depicted here is to them the manifestation of a whole society undergoing a monumental catharsis, to purify itself of all its sins. A few would probably agree that pure erotica was only a logical conclusion to the theme of "love for all aspects of life," sex being an undeniably important "aspect" of life. It is redundant to add that the Brahmin keeper of the temple was reluctant to be left out of any "aspect" of life, particularly of course of one as immortal as sex. So why should not the walls of his temple become inviting visual lessons in the irrepressible art of sexual enjoyment? The cynics have even unabashedly suggested that the Brahmin having earned his due from performing the marriage ceremony posted erotica on his temples to ensure continued indulgence by the novice couple in affairs of the temple. That the Brahmin judged human nature with astute shrewdness is more than amply proven by the worldwide undying fame of the erotica of Khajuraho.

Whatever the causes, there is little doubt that in the display of excessive erotica, the Indian sculptor carried out his master's bidding with, if anything, an even greater skill and sense of plastic beauty (*Fig 18.11*) than he lavished on statues of the gods themselves.

➤ *Pages 190 and 191*
Figs 18.11 Erotica at (a) the Sun temple, Konarak and (b) on the base of the Kandarya Mahadeo temple at Khajuraho

Fig 18.11a

Fig 18.11b

CHAPTER 19

Cholas and Pandyas—The Magnificent Tamils

In the South too, as in the North, the tenth and eleventh centuries constituted the age of 'temple cathedrals.' As the great Chola dynasty began to dominate the southern country, the refined little gems of the Pallava period were no longer an adequate expression of its majestic power. The Cholas desired something grander in conception to immortalise their glory. The craftsmen having acquired long experience in the skills of stone masonry, were more than qualified to carry out their masters' grandiose schemes.

Great Brihedesvara Temple, Tanjore

Once Thanjavur (modern Tanjore) fell to the Chola king Rajaraja the Great (985–1081), he commissioned the Brihedesvara temple (*Fig 19.01*), the size alone of which was a gigantic step forward in the evolution of religious architecture in the South. Until now, even after 500 years of building experience, no tower higher than 60 ft (18.2 m) had been attempted by the Hindu craftsmen of India. But the masons of Tanjore now set out to make amends for the timidity of the southern builders in one big leap, conceiving of a spire more than three times the maximum height achieved so far. An ambitious superstructure required that the cellas be proportionately enlarged in plan (*Fig 19.02*). In the great temple of Tanjore, it became a massive square of 82 ft (24.⌐ m) side, containing within it a *pradakshinapath*, running all the way around the cella. The cube of the *garbhagriha* was carried vertically up to a height of 50 ft (15 m). Thereafter, the familiar pyramidical tower soared another 130 ft (39.6 m) into the sky, capped by a single enormous domical stone weighing more than 80 tons.

Two flat-roofed *mandapas* in front of the *vimana* are placed as always along the central axis, the largeness of their size in keeping with the proportions of the *vimana*. The portico housing the sculpture of the holy Nandi bull is situated along the same axis but is detached from the main temple, much like the *sabha mandapa* of the Sun Temple at Modhera and the Nat Mandir of Konarak. The entire complex stands today in the middle of a quadrangle defined by a peripheral *veranda* of two rows of columns.

The 200 ft (60.8 m) high stone tower (*Fig 19.03*) which rises impressively over the sea of courtyards around, is a gigantic reproduction of the familiar Dravidian *vimana*. However, the horizontal tiers of its thirteen storeys have been suppressed to accentuate the soaring verticality of the converging lines of the truncated pyramid. The *vimana*, though only a few feet less in height than the projected height of the *shikhara* of Konarak, and almost as old, has withstood the ravages of over 900 years, literally without a dent. In constructing it the south Indian craftsman showed a greater understanding of structural principles. It may be argued that in choosing the simpler and more stable form of the pyramid he

◄ *Facing page*
Fig 19.01 The over 55-metre high vimana of the 10th century Brihideswara at Tanjore capped by a single-piece domical stone weighing over 80 tonnes

Λ
Fig 19.03 The vimana of the great temple at Tanjore over the courtyards around

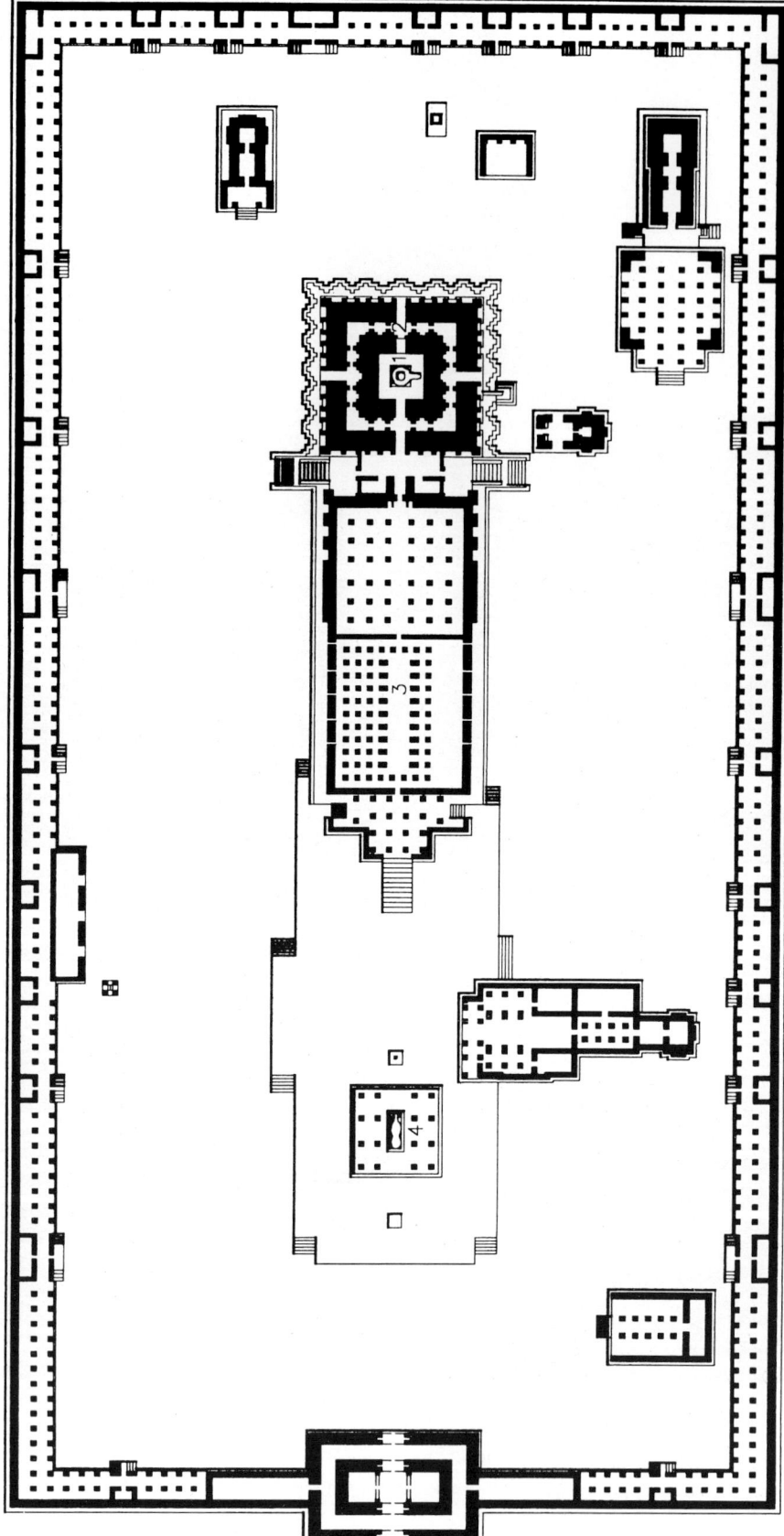

Fig 19.02 Plan of the great
Brihidesvara temple, Tanjore.
1 Garbhagriha, 2 Pradakshina,
3 Mahamandapa and 4 Nandi

was on safer grounds than the northern builder struggling with the more complex curvilinear shape of the *shikhara*. But this would not be entirely true; as we shall see, he soon proved himself equally adept at constructing structures of equally complicated nature, endowed with enduring stability.

Expansionism of the Cholas

Rajaraja the Great was followed by his son Rajendra I who continued the expansionist policies of his father. His ambitions, however, were not restricted merely to subduing the neighbouring Chalukyas who were in control of the trading ports of modern Malabar. He moved much further afield and sent a Chola expedition north-east across Orissa which is said to have reached the holy waters of the distant river Ganga. The Cholas were also engaged in trading activities with China and to safeguard their shipping interests, they waged a successful campaign against the kingdom of Shrivijaya (the southern Malay peninsula and Sumatra). Closer at home, the threat posed by the triangular alliance of Kerala, Sri Lanka and the Pandyas, was also crushed by Chola pressure. The Cholas eventually overran and devastated Anuradhapura, the capital city of Sri Lanka, and set up a new capital at Pollonnarava.

The Temple of Gangaikondacholapuram

Fig 19.04 The fluid lines of the vimana of the Great Chola temple at Gangaikonda-cholapuram (AD 1025)

Growing Chola power was accompanied by an equally passionate desire to build new cities that would be an appropriate architectural symbol of their expanding empire. Kingly ambitions were reflected in Rajendra I's decision to challenge the glory of Thanjavur by shifting his seat of power to Gangaikondacholapuram (28 km from modern Kumbhakonam). The spirit of the times dictated that the most dominant edifice of the new city be a monumental temple. The greatness of an Indian city of the middle ages was measured not by its town planning wonders, nor its streets and bazaars and houses, but rather by the magnificence of the temple housing the presiding deity. Today, though, all that the deity of Gangaikondacholapuram presides over is a few straggling village huts, Rajendra's great city having long ago crumbled into dust. The city temple, though reduced to ruins, is a flamboyant reminder of vanished Chola glory.

Though the architects of King Rajendra obviously modelled their temple in essence of design as well as scale on the great temple of Tanjore, they made some refreshing innovations nevertheless. The main *vimana* (*Fig 19.04*) rising from a base even larger than that at Tanjore, is transformed into a form embodying concave quoins, at the ends of convex planes of sculptured surfaces. In contrast to "the grid and geometrically perfect *vimana* of Tanjore that symbolizes conscious might, the fluid lines of the latter are imbued with a subconscious grace." The *mandapa* of this temple on the other hand is totally different from the concept of Thanjavur. It is a huge flat-roofed hall of 75 × 195 ft (22.8 × 59.4 m). The entire structure is built on a high platform. The approach aisles designed like processional paths, however, are left at ground level. With the Hindu builder's aversion to any system of construction other than the trabeate, it was inevitable that such a large room be held up by a veritable forest of over 150 closely spaced, slender columns (*Fig 19.05*). The first of its kind, this hall foreshadows the emergence of the famous halls of thousand columns of later temple complexes.

Fig 19.05 Plan (above) and section (below) of the Gangaikondacholapuram temple. 1 Garbhagriha, 2 Mahamandapa, 3 Mandapa, and 4 Nandi

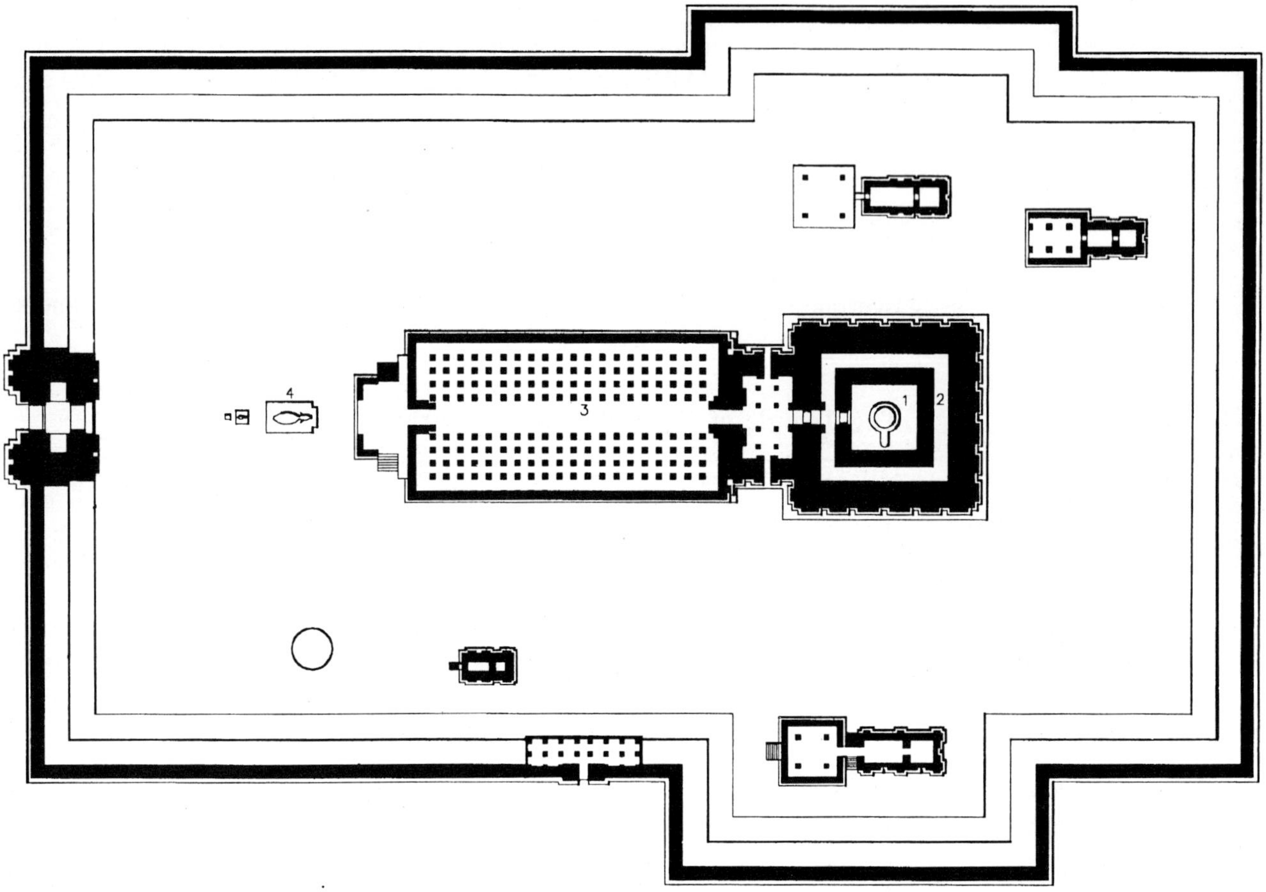

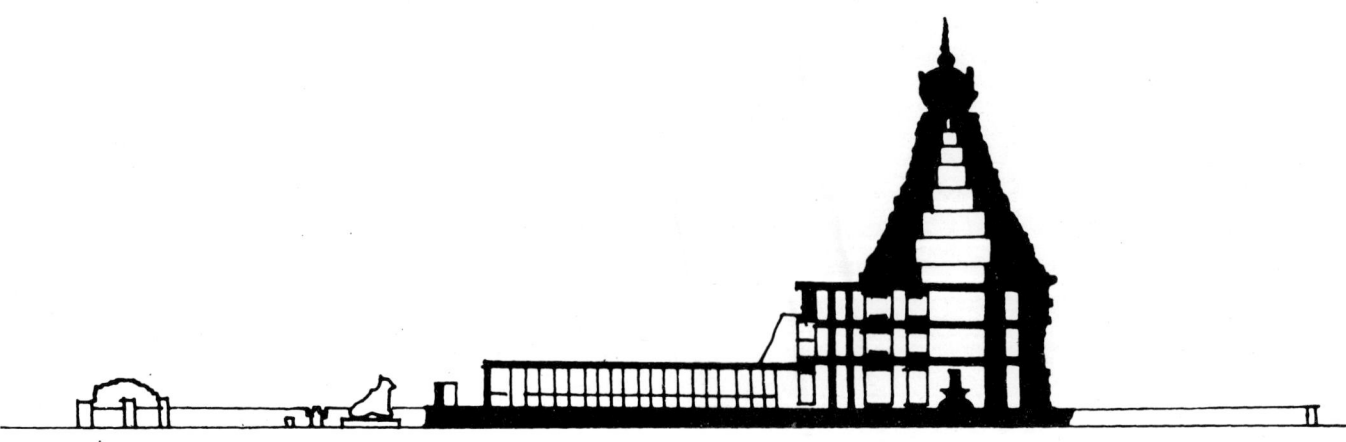

PLAN

SECTION

The Triumph of the Pandyas

With the death of Rajendra I in about AD 1050 his successors continued to maintain communications with the Chinese to sustain their commerce. King Kulothunga even sent an embassy of seventy-two merchants to negotiate trade agreements with China. Nearer home, their peninsular neighbours, however, were forces to contend with. Even though the Cholas sacked the Chalukyan capital of Kalyani, they were unable to decimate the seemingly indefatigable Chalukyans. Rather, the rivalry between the two was only further intensified. The southern triumvirate of the Pandyas, Keralas and Sri Lanka, too, never gave the later Cholas sufficiently long periods of respite to indulge in building ventures on the Tanjore scale. Gradually, sapped of their strength after 800 years of indisputable sway over the Tamil country, the Cholas ultimately succumbed to the Pandyas in the middle of the fourteenth century.

The Pandyas inherited both the wealth of the trade that the Cholas had established with the Chinese and the splendour of their trading cities. The kings continued to live in great luxury and magnificence. Marco Polo, passing through India on his return from China, described the port town of Kayal (today marked by a few fishermen's huts on the beach) as a "great and noble city" and the king as one who "wore upon his person the most costly jewels." By the times of the Pandyas, however, dramatic changes had taken place in the polity of the North. A rather incomprehensible force had crossed over through the Khyber Pass and established domination in the northern regions. This was the armies of the followers of a new and unvanquished religion—that of Islam. They had by now established their capital city at Delhi and were beginning to make inroads into the Hindu empires of the Deccan.

The Sanctity of the Ancient

As during the later Chola period, construction activity, as far as new massive undertakings were concerned, continued to languish during Pandyan suzerainty, though for different reasons. While the temples of Tanjore and Gangaikondacholapuram, adequately glorified the power of the Hindu gods as well as the Chola kings, they never became popular centres of worship. The religious sentiment of the people was directed more towards ancient temples in spite of the fact that these were modest in size and possessed little architectural significance. To the devotee, however, old was gold, and to him these historic shrines were endowed with "marked sanctity because enshrined within them were images of deep and lasting veneration." To replace or materially change the outer shell of these ancient consecrated images merely to enlarge them in size, would have been a sacrilegious act. And so, a good deal of the energies of the building craftsmen under the Pandyas was concentrated in the maintenance, addition of ancillaries, and the enlargement and protection of such ancient structures.

Battlements of Protection

To guard their immense wealth, high walls of a purely utilitarian character, often battlemented and provided with platforms were hurriedly built around a number of such sacred temples. Such activity was, no doubt, prompted also by the imminent danger of Muslim invasion from the North. The infamous sack of Somnath and the daring raids of Allaudin right up to Deogiri were dreaded and fresh memories. The external appearance of many an ancient shrine was thus forcibly reduced to the plainness of a fortress—a function that the temple could actually be called upon to fulfil in an emergency.

Gateways to Temples

This state of affairs, however, could not be allowed to last for long. The problem facing the royal patrons was of adding architectural status to the insignificant ancient shrines lost within its protective battlements. Finally, the Pandyan designers hit upon the idea of erecting lofty gateways at the entrance points in the walls. This served the two-fold object of alleviating the monotonous drabness of the enclosure, and at the same time endowing the temple precincts with appropriate visual impact. The Pandyan builders were now called in to erect these monumental passageways. During subsequent periods of peace and security, lofty portals or *gopurams* (*Fig 19.06*) (literally, the cow-gate, from rural Aryan terminology) could be erected in the outer walls, without disrupting the ritual or endangering the security of the temple.

◄
Fig 19.06 Section of a typical Southern gopuram

The Evolution of the Gopuram

Inevitably, an entrance to the house of God had to be a massive form towering into the sky and visible for miles around. Yet it was undesirable to rival or repeat the architectonics of the *vimana* of the inner cella, the holy of holies. In any case, the square plan of the *garbhagriha* was hardly an appropriate one for a gateway; a square apart from its sacred sanctity, had a sense of firmness and finality about it. On the other hand, the ideal design of an entrance portcullis would be one that conveyed a sense of transition from outside towards the inner sanctum. A broad side on rectangular plan conveyed this intention more appropriately than the square (*Fig 19.07a*). Thus, the base of the gateway became a vertical pile of masonry rising over the oblong of the plan. The central opening was spanned by means of massive lintels of stone. If the size demanded, a system of projecting brackets was inserted on either side, to reduce the span of the lintels (*Fig 19.07b*).

Over this substantial base, built in stone, rose a pyramidal structure composed as ever of diminishing tiers generally built in brick and plaster. The topmost tier, like the base from which it had emerged, was also a rectangle in plan. It would thus not be appropriate to crown it with the familiar octagonal domical stone, which sat more comfortably over a square plan. Moreover, such a symbol, associated as it was with the sacred tower over the central shrine, belonged to a higher religious hierarchy. Looking back to the wide array of religious and secular forms of the past, for an adequate alternative the craftsmen found that the barrel-vault roof of the rectangular Buddhist *chaitya* had served his purpose eminently. This functional form, reduced now to an ornamental solid shape, crowned with a row of finials over the ridge, proved a fitting climax for his soaring tower (*Fig 19.07c*). Every inch of the surface of the body of the *gopuram* was covered, of course, with the usual efflorescence of tier upon tier of sculptured and even polychromed gods, *apsaras* and demons. Over a period of time, a number of such towers rising around a temple became the familiar skyline of any sizable urban settlement in the southern country (*Fig 19.08*). The temple had by now evolved into the centre not only of the religious but also social, economic and cultural life

Figs 19.07 (a)–(c) The evolution of the form of the gopuram, derived as it was from a rectangular prismatic base, and a pyramid crowned with a barrel-vaulted form derived from the Buddhist chaitya ➤

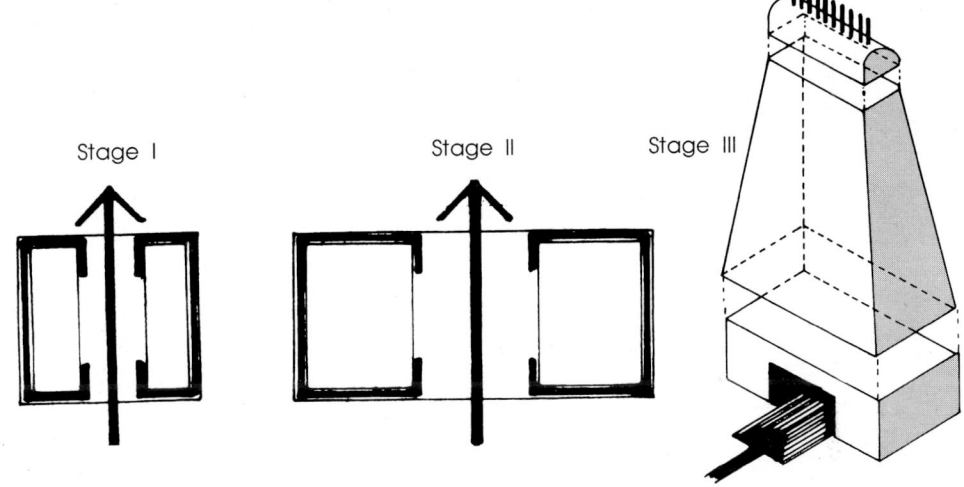

Stage I   Stage II   Stage III

Fig 19.08 A number of gopurams placed along the cardinal axis of a South Indian temple ⋎

of the urbanites. The maintenance and financing of this social centre was comparable to that of any large-scale institution of modern times. In its heyday, the great temple of Tanjore, for example, had an "annual income of 500 lb troy of precious stones and 600 lb troy of silver. It kept in reasonable comfort an army of 212 attendants, 57 musicians and 400 women entertainers, apart from the many hundreds of priests."

Sociology of the City Temple

Temples also generated a steady income by financing commercial enterprises and acting as bankers and moneylenders. Another perpetual source of income was the so-called *devadasis*. Originally, they may well have been a venerated group of women attendants donated by wealthy parents at an early age to the temple, not unlike the vestal virgins of Rome. Some of them were painstakingly trained by the learned priests in the art of the South Indian classical dance form Bharatnatyam. A larger number of them, however, were reduced to prostitution operating under a religious umbrella and thus enriching the coffers of the temples.

Such a phenomenal growth in the functions of the temple naturally demanded a corresponding physical growth. However, having enclosed their temple precincts within a forbidding wall, the only way to acquire more land for the complex was to throw another concentric ring of even larger battlements, punctuated by equally huge *gopurams*. The seeds of the growth of a temple akin to the annual rings around the trunk of a tree were thus firmly sown. As we shall see, the germ of this idea led to the evolution of the great temple cities of Madurai and Rameswaram. In building such temple cities in the South, the builders were kept occupied by what contemporary architectural parlance would describe as 'addition and alteration jobs'.

CHAPTER 20

The Hoysalas of the Karnataka

In the north-western reaches of the peninsula, meanwhile, an entirely new form of building activity was beginning. The Muslim rulers of the Delhi Sultanate having established their power over most of the Gangetic plain, were now steadily pushing southwards. A good number of craftsmen, concerned less with religious loyalties and more with exploiting their professional and technical skill, were absorbed into the construction of edifices for the Muslim rulers. Temple building activity of any consequence was virtually at an end in the North and the more orthodox master craftsmen were immigrating southwards to seek the patronage of the surviving Hindu courts.

The northern Deccan region between the river Tapti and the river Krishna was controlled by the Yadavas who had thrown off the Chalukya yoke. They were not averse to employing these craftsmen to promote a building style distinct from that of their former overlords. The craftsmen, seeking refuge in the South were descendants of those who had evolved the Gujarat style under the once powerful Solanki dynasty. Architecturally, the Yadava country gradually became a provincial outpost of Gujarat. The thirteenth century temples of this region do show some signs of originality, but in substance they are largely variations on the themes of Gujarat.

An example of this is the Ambernath Temple (*Fig 20.01*) at modern Thane near Bombay. In plan, it follows the familiar arrangement of the *mandapa* and the cella placed in a diagonal relationship—a planning device that was popular under the Solankis. In the true Gujarat tradition, the craftsmen have really gone to town in decorating the walls and ceilings of the temples with finely crafted details. The wealth of sculptural figures and the series of projections and recesses are piled up with "ornament upon ornament producing a bewildering tumult of sculptured forms." The tumult continues even into the roof of the *mandapa*, which is a pyramidical pile of miniature reproductions of its own architectural form (*Fig 20.02*). Though the Ambernath, like so many other temples in Gujarat, has lost its *shikhara*, the temple at Sunnar in Nasik is better preserved. It is a temple of the *panchtayana* class, standing in the middle of a wide platform with miniature shrines at each corner. The tower over the central shrine is a series of full-bodied *shikhara*, like forms.

The Hoysalas and Northern Influence

Soon, however, with the ominous shadow of Muslim power spreading further, the Hindu craftsman was pushed deeper south into the region around modern Mysore. Here a tribe of hill chieftains known as the Hoysalas was emerging into power after a long struggle with the once omnipotent Cholas. The bitterness of their battle could not have left the Hoysalas with great love or admiration for the architectural style of their arch foes. Racially, too, the Hoysalas had greater affinity with the North and the Chola buildings held no orthodox sanctity for them. Rather, the north Indian

Fig 20.01 The eleventh century Ambarnath temple at Thane near Bombay; plan half looking up and half looking down, showing the elaborate geometry of floor and ceiling patterns

Fig 20.02 The pile up of minishikharas in the Ambarnath temple

craftsman under such conditions was more than welcome to make his contribution to the evolution of a new style. However, faced with the equally strong traditions of his Dravidian counterparts, he was not going to be able to freely erect direct copies of *shikharas* and *mandapas* as he had for the Yadavas in the Deccan. His contribution perforce had to be of a muted and subtle nature. This fortuitous meeting of master craftsmen brought up in the traditions of the two great styles of Hindu architecture, the northern and the southern, resulted in a third style, classified as that of the Hoysalas. There must have been a fine spirit of cooperation between the masters, for the style that emerged is neither mere plagiarism of conflicting symbols, nor an unseemly clash of two styles. Rather, features of the two are subtly merged together into novel forms that have a distinct and refreshing flavour all their own. The wheel had beautifully come full circle. It was this very region that more than 700 years ago under the Chalukyas had seen the birth literally side by side of the pre-eminent features of the northern and southern styles, the *shikhara* and the *vimana*. Now the builders were back at it again evolving yet another style by the merging together of what may be called the two mother styles.

The Temples of Ittagi and Gadag

The synthesis that emerged is seen in its early experimental forms on the fringes of the Hoysala kingdom which was centred around ancient Dwarsamudra in modern Mysore. It is obvious from even a glance at the various temples at Ittagi, Gadag and Lakhundi that the northern craftsman's most conspicuous contribution to the synthesis was the change he wrought in the form of the tower over the cella. The marked horizontal tiers of the tower were submerged into a profile, imbued with some of the sensuous grace of the north Indian *shikhara*. The numerous recesses and chases of the base were carried up into the tower of the newly evolved 'vimana shikhara' giving it in plan a form more curvilinear than the *vimana* and yet not quite the pronounced parabolic profile of the *shikhara* (*Fig 20.03*). The typical circular shafts of the columns of these temples were turned out from a

Fig 20.03 A typical vimana shikhara of the Ittagi Dharwar style

stone lathe, surely suggested by the craftsmen who had moulded similar ones for the *mandapas* of Gujarat. The repeated application of the motif of the Indo-Aryan *shikhara* on the walls of the temples is rather forced and incongruous. Gradually, this hybrid development of Ittagi and Gadag matured into an identifiable style, challenging the pristine beauty of either of its generic sources. Soon it was acceptable even to the imperial Hoysala rulers at their capital city Dwarasamudra, who perceived in it the potential to rival the more orthodox southern building style of the Cholas and Pandyas.

The Famous Star-Shaped Temples

To highlight the distinctiveness of their temples, the architects, with due encouragement of the Hoysala kings, flamboyantly set about evolving entirely new patterns for their places of worship. First, they planned temples and shrines within which were situated more than one central *garbhagriha* or inner cella. These cella, even up to five in number in some temples, were grouped along one end of a large common *mandapa*. Further, the plan of each of these cellas was an elaborate star shape, seemingly defying the sanctity invested in the square by the Hindu art canons. On analysis, however, one finds that the builders had derived even their complex stellar outline from the purity of the square (*Fig 20.04*). This was achieved by rotating the square around its fixed centre, and turning its diagonal through a series of equal angles. The resulting outline of the interesting corners of so many superimposed squares merges as a star. The number of its points and their proportions could be varied merely by changing the angle through which the diagonal was turned at every step. By this ingenious device the artist paid due technical deference to the orthodoxy of the square and, at the same time, added a new dimension to temple planning. He was obviously so pleased with his innovation that even the platform over which the temple was raised, echoed the outline of the cella. The towers crowning the various cellas acquired a bell-like profile which was a streamlined version of the *shikhara-vimana* that had been evolved by the builders of Ittagi and Dharwar. The *mandapa* remained the familiar cross-shaped pillared hall. It was distinguished, however, by the circular shafts of the columns, which now emerge straight from the artistic hands of the lathe cutter, adorned with a series of parallel, almost knife-like edges of various shapes.

Fig 20.04 Revolutions of the square around its centre give rise to a star like profile

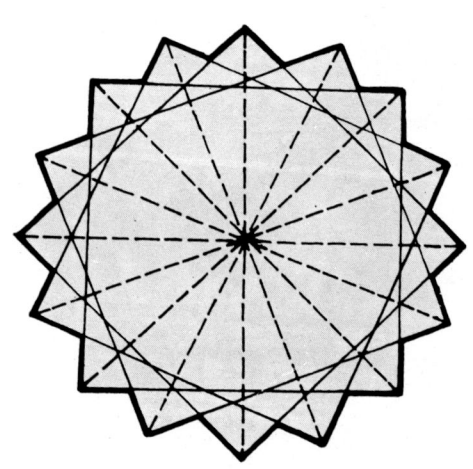

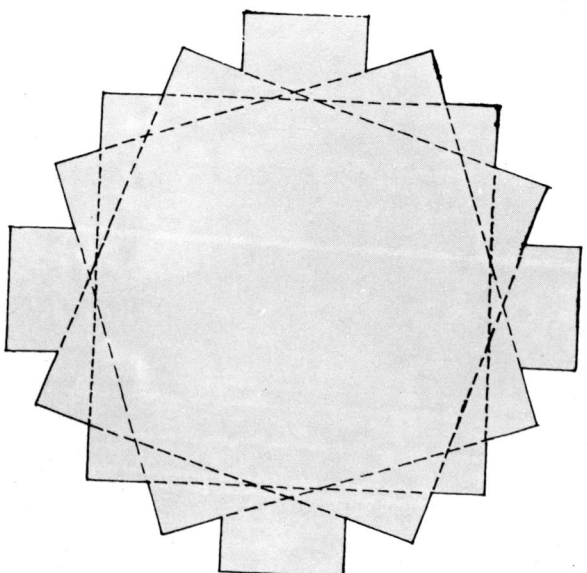

Temples of Ductile Chlorite Schist

To build temples of such complexity with unwieldy blocks of sandstone would have been a formidable task indeed. However, a material easier to work with was readily available—a closely textured chlorite schist which was soft when quarried, but turned to adamant hardness on exposure to air. This special quality made it the ideal medium for the expression of the elaborate architectonics of the Hoysala builders. The cutters turned blocks of the elaborate schist on the lathe to reduce them to circular shafts. These were invested with a profusion of concave and convex mouldings with almost knife-like edges. The squat horizontality so acquired by these columns (*Fig 20.05*) became the theme of the entire elaborate decorative scheme, from the stylobate of the base of the apex of the bell-like tower.

Fig 20.05 The knife-like edges and the horizontal delineation of the columns of the Hoysala temples

The Hoysala king Vishnuwardhan is credited with the rebuilding of the old capital of Dwarasamudra, naming it Halebid (*hale*—old, and *bidu*—capital). He is reputed to have initially commissioned his chief architect, the legendary Janaka Acharya, to build a temple to the god Channa Keshava (*Fig 20.06*) at Velapura (modern Belur). Though the temple is built around a single shrine, Janaka Acharya incorporated into it all the other essentials of the Hoysala style—star-shaped shrine and platform, bell-shaped towers, and pillared *mandapas* (*Fig 20.07*). Each of the twenty-eight openings between the pillars of the *mandapa* was later closed up with a trellis work of a rather incongruous design. This was meant to provide privacy for the Brahmins watching the rituals, particularly the highly seductive dancing of the *devadasis*. The *mandapa* of this temple has one of the most beautiful circular stone platforms. Over the centuries it has acquired a lustrous polish from the incessant dancing feet of the exponents of Bharatanatyam. The entire temple is set within a vast rectangular enclosure of 440 × 360 ft (134 × 110 m) to which the traditional *gopurams* were added at a later date.

Fig 20.06 Plan view of the twelfth century Channa Keshava at Belur

The City of Halebid

Carried away by his great religious fervour and immense faith in Janaka Acharya, the same king commissioned an even more elaborate form of the Belur temple. This, the great Hoysalasvara temple, became the focus of the capital city of Halebid. It is a temple even larger and richer than its predecessor at Belur. The architect proceeded to duplicate his effort at Belur by laying out (*Fig 20.08*) at Halebid two identical temples parallel to each other, connected only at their transepts. Each temple has, in addition, its own Nandi *mandapa*, detached from the main body of the shrine.

Fig 20.08 A view of the, great temple at Halebid (AD 1150)

Fig 20.09 Side elevation of the temple at Halebid

Sculptural Architectonics of Janak Acharya

Here, at Halebid, the master Indian sculptor, working with the tactile material of the chlorite schist, carved out the most fitting climax to the centuries of his art in India. Virtually not an inch of this vast composition escaped his chisel. The walls of the Hoyselasvara suggest the "analogy of a voluminous, constantly unrolling illustrated scroll," the entire elevation (*Fig 20.09*) being a "scheme of closely crafted continuous mouldings, borders, friezes, cornices and bands of statuary carried all around the building." The richness of the sculpture is so awe-inspiring that one scarcely notices the lack of the bell-shaped towers over the cellas. These were either never completed or having been constructed around a core of brick were easily desecrated later. Profusely covered with such masterpieces of sculpture as it is, the "Halebid temple and the Parthenon are probably the two extremes of the architectural art of the world." The one revels in the cold purity of its architectural form, and the other in the warm complexity of its sculptural architectonics.

Kesava at Somnathpur

The most complete extant example of the Hoysala builder's art is, however, the Prasanna Channa Kesava in Somnathpur (*Fig 20.10*) near modern Bangalore, built by Vardhana in AD 1270. Set inside a rectangular courtyard, the temple consists of a symmetrical arrangement of triple shrines at the end of a rectangular *mandapa* and *navranga* (being only another regional term for a dancing hall). It is the only example of Hoysala art that has retained its three bell-like towers in a well preserved form.

Not entirely content with a stellar plan for the cellas, later builders attempted to transform the plan of the *navranga mandapa* also into the shape of a star in the Dumbal temple at Dosappa. The first experiment could not be elaborated upon, since the collapse of the Hoysala dynasty in AD 1326 cried halt to the work of the Hindu craftsmen in the Karnataka. Even if the Muslim invasion of the South had not eroded the power of the Hoysalas, whether the Hindu craftsman could possibly have excelled himself beyond the "prodigal plastic manifestations of his art" at Halebid, Belur and Somnathpur, is indeed a moot point.

Fig 20.10 (a) The bell-shaped tower and (b) plan of the Prasana Channa Keshava at Somnathpur

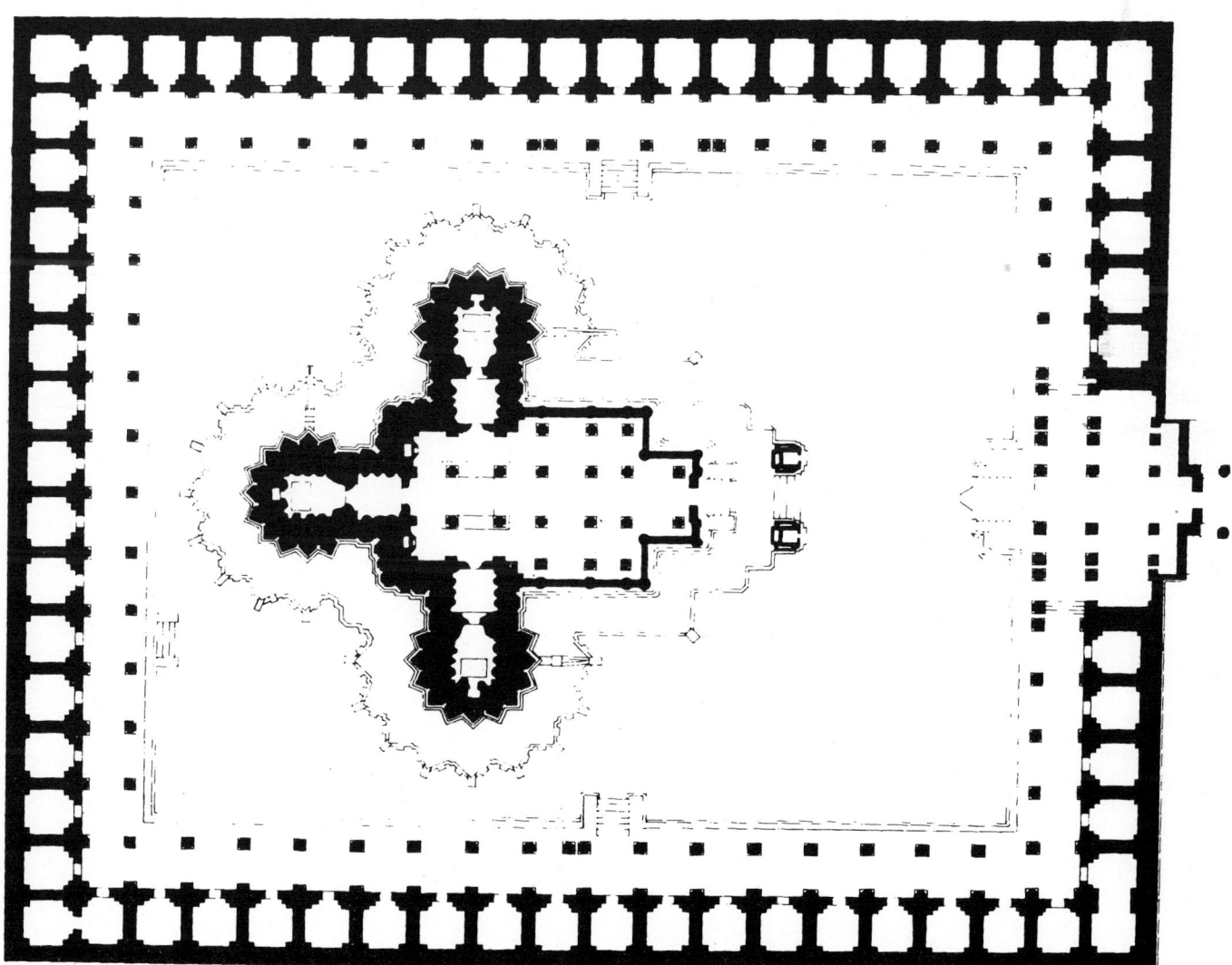

CHAPTER 21

Vijayanagar—The Last Great Glory

The Sultans of Delhi, content with their nominal conquest of the Deccan, established a Turkish Governor at Daulatabad and retired to more urgent imperial business in their capital city of Delhi. The Governor, true to tradition, followed the familiar pattern of revolt, and at the earliest opportunity proclaimed himself King Bahman Shah, with his capital at Gulbarga in modern Mysore. He was, however, unable to expand his empire in the South since the two Hindu kingdoms—one centred around Warangal (modern Andhra) and the other at Hastnavati (modern Hampi)—were not altogether obliging. Ultimately, the Sultans sent armies from Delhi to reclaim their kingdom in the South. In the process they carried back two Warangal princes, Harihara and Bukka, to Delhi. The princes were persuaded to convert to Islam and sent back, charged with the mission of establishing the Sultan's authority.

Harihara astutely got himself crowned King of Hampi and was even accepted back into the Hindu fold. He proved to have been well schooled in the arts of intrigue at the Muslim courts of Delhi. Ably aided by Bukka and proclaiming independence from his Muslim overlords, he went about the task of consolidating his newly acquired empire. A new empire, of course, could be proclaimed only through a new capital city. For strategic reasons, Harihara decided to expand the existing town of Hampi, situated though it was on "a tossed and tumbled jungle of sunblasted stones with only limited open spaces intervening." Rather, he took advantage of the ready availability of material from the very site of construction as it were, and completed the first phase of expansion in a short period of seven years.

Coast to Coast Empire

The empire founded by Harihara continued to grow. Within a period of a hundred years it stretched from the eastern to the western sea coast. Such a vast domain was defended by an enormous army "comprising a million fighting troops in which are included 35,000 cavalry in armour." Curiously enough, though horses were the mainstay of such a force, the Indians never took seriously to horse breeding. They purchased them from Portuguese traders who had by now entrenched themselves in Goa. The Portuguese, apart from trading "Arab horses, velvets, damasks and satins" for "jewels and pagodas which be ducats of gold" had carefully established friendly relations with Hindu Vijayanagar, the Muslims being their major rivals for control of the trade across the Arabian Sea. Hampi, meanwhile, continued to prosper, and in keeping with the series of conquests and victories grew into the famous Vijayanagar— the city of victory.

By the time of King Devaraja II, less than 100 years after its founding by Harihara, Nicolo Conti, an Italian traveller visited Vijayanagar in 1420. He estimated (somewhat exaggeratedly it would seem), that "the circumference of the city be sixty miles." He was "much impressed by the strength of the fortifications which were carried up the hills so as to enclose the valleys at their base." These valleys were watered by constructing a huge dam in the Tungabhadra river, which conveyed water to the city through an aqueduct fifteen miles in length, cut out of the solid rock for a distance of several miles." The outlying areas of the "tossed and tumbled jungle" were transformed into cultivated fields and pastures, well stocked with flocks and herds. "The city was therefore so well supplied that it could stand a siege for an almost indefinite period."

The Marvels and Blandishments of Vijayanagar

The strength of Vijayanagar is amply borne out by the remains of the fortifications of the city; we can see that the builders had left little to chance. They have been described by a contemporary Abdur Razzaq, an ambassador sent by Sultan Shahrukh of Herat to the Zamorin of Calicut. After travelling across "streets as wide as tourneys" he found that Vijayanagar was "so built that it has seven fortified walls, one within the other. The seventh fortress is placed in the centre of the others and occupies ground ten times greater than the chief market of Herat. In that is situated the palace of the king. From the northern gate of the outer fortress to the southern gate is a distance of two statue *parasangs* (about 7 or 8 miles), and the same with respect to the distance between the eastern and western gates. From the third to the seventh fortress, shops and bazaars are closely crowded together."

> *Fig 21.01 The so-called Soolai Bazar, one of the remains of the once grand city of Vijayanagar of the sixteenth century AD*

The best of the bazaars (*Fig 21.01*), true to the tradition of a thriving Indian city of the middle ages, were the houses of the *devadasis*, "the (euphemistic) handmaids of the gods." The prostitutes' quarter, in fact, was "one of the sights of the capital." The splendour of their houses and the beauty of the hearts ravishers, their blandishment and ogles" were to Razak "beyond description." Apart from providing for the sexual appetites of the inhabitants of Vijayanagar the tax on the proceeds of the brothels, if one is to believe Razak, paid for 2000 policemen whose salaries amounted to 12,000 *pafanams* daily.

The Palace of Krishnadevaraja

Under the rule of King Krishnadevaraja, a contemporary of King Henry VIII, the city was visited by Pae, a Portuguese traveller to whom it seemed "as large as Rome and very beautiful to the sight." It consisted of over 100,000 houses and must at the time have had a population of over half a million. The king's palace "stood in an enclosure that contained thirty-four streets in the heart of the city. The private apartments of the palace were in a separate building known as the house of victory. One of them was panelled with ivory from top to bottom. The pillars were made of rosewood decorated with flowers and lotuses, all of ivory, and well executed, so that there could not be better."

The king's audience chamber was elevated above the rest. "In this agreeable locality one sees numerous running streams and canals formed of chiselled stone, polished and smooth." The wilderness that the city is today was once covered with "many lime and orange trees, growing so closely together, one to another, that it appears like a thick forest."

Finance and Religion

Such a magnificent Hindu city could not conceivably have existed without the support of the priestly class. King Krishnadevaraja of the sixteenth century, under whom Vijayanagar reached the climax of its glory, conferred "almost fabulous wealth as endowments on temples and Brahmins." This ensured that the Brahmins would not challenge the rather questionable heredity of the kings from a Muslim convert.

A large part of the endowments went into financing the ever expanding ceremonials of religion. The rituals had become so elaborate that it was no longer possible to contain them within the traditional axially aligned compartments of a temple. The temples of Vijayanagar, thus, are not the unified compositions of the Cholas or the Hoysalas. They are, in fact, a number of small units, each with its own prescribed function, set rather haphazardly as dictated by the terrain, within a large rectangular enclosure. The basic unit of such a temple conglomerate was a flat-roofed hall, supported on a series of elaborately crafted columns.

Nature and Abodes of God

In the temples and palaces of Vijayanagar, often the living rock itself became the foundation of a grand superstructure and "it is sometimes difficult to tell where nature ends and (man-made) art begins" (*Fig 21.02*). The builders have, unintentionally perhaps, left behind a most graphic record of perfect harmony between art and nature. "The striking, almost planet-like landscape of the region," lives here with the handiwork of man, as if the two were never separate.

Fig 21.02 In Vijayanagar, it is some times difficult to tell where nature ends and art begins as in this, the Raghunath temple, built virtually over living rock ⅄

The Great Vithala Temple

The Vithala temple (*Fig 21.03*) commissioned by Krishanadevaraja as late as the beginning of the sixteenth century, is spread over an area of approximately 500 × 300 ft (152 × 91 m). The central part is a series of conjoined pillared halls extending horizontally over a distance of 200 ft (61 m) but nowhere rising more than 25 ft (7.6 m). The tower over the *garbhagriha*, if one was ever erected, has vanished. It was most likely constructed in brick and plaster. Beyond the extensiveness of its conception there is little to commend it, since the temple is devoid of any tangible architectural form. The Hindu craftsmen of Vijayanagar concentrated their energies on the myriads of columns, each of these becoming a miniature architectural composition in itself (*Fig 21.04*). Interposed between the columns is a "half natural, half mythical" relief of a lion, resurrected like a ghost from the ancient Pallava period. Each pillar expands below into a massive pedestal and above into grotesque brackets of gigantic proportions. Any concept of space in the flat-roofed halls is devoured by clusters of such columns. Ultimately the halls produce "an effect of bewildering intricacy" rather than comprehensible art. Within the Kalyana *mandapa*, attached to the temple, the marriage of gods and goddesses was conducted annually. It is, if one can imagine it to be possible, even more elaborately wrought than the main *mandapa*. Capped by the same "voluptuous double flexured cornice with a turned-up outer edge" this hall is held up by twelve piers around a square throne in the centre. The major sculptural relief that stands out from this forestry of columns is an imitation chariot car (*Fig 21.05*) which is wrought from a single block of hard granite. Its stone wheels, lifted a few inches from the ground, actually revolve around their axles.

Fig 21.03 The kalyana mandapa of the Vithala temple capped with the typical double flexured cornice of the Vijayanagar style

*Fig 21.04 The myriads of
columns in the hall of the
Vithala temple at Vijayanagar*

Temple of Hazari Ram

The same king also commissioned the building of a private chapel, for some curious reason now called the temple of Hazari Ram (*Fig 21.06*). Architecturally, it is an undistinguished as the Vithala, but the sculpture "attained an extravagance that has an inevitable suggestion of the grotesque and fanciful." This effect is highlighted by the Amman shrine at the rear where, for good measure it would seem, the sculptor added the keel root form of the Buddhist *chaitya* hall to the ever expanding chiaroscuro of fanciful shapes.

It would seem as if Hindu architecture in its dying glory was momentarily but rapidly improvising on scenes and memories of its long and distinguished past; a glory that was finally shattered in a great crescendo of violent upheaval. Krishnadevaraja had been succeeded by his son Achyutarya who succumbed to the machinations of his minister Ramaraja. Ramaraja secured the throne in the name of Krishnadevaraja's son, and was for a while successful in his intrigues with the Muslim kings of the Deccan. Due to his growing arrogance, however, the Muslims ultimately formed an alliance to challenge his power.

Fig 21.06 The Hazari Ram
temple at Vijayanagar

Intrigue and Islamic Triumph

On the banks of the Krishna, 25 miles from the town of Talikot, the 600 guns of the Muslim confederation routed the Vijayanagar army of between half a million men, besides a multitude of elephants. The Muslim army, after a day's rest attacked the capital city and, for a space of five months, Vijayanagar knew no rest. "Never perhaps in the history of the world had such havoc been wrought on so splendid a city. The magnificent stone carving was smashed to pieces with crowbars and hammers and where it defied human efforts fires were lit to burst it open. Teeming with a wealthy and industrious population full of plenitude and prosperity one day—the proud capital was (soon) a forlorn ruin inhabited only by tigers and other wild beasts." It was reduced to what Percy Brown 300 years later described as "the striking and almost planet-like landscape."

The empire never quite recovered from the disastrous battle of Talikot. Ramaraja's unfortunate successors were left to rule a sadly diminished kingdom, from Penukoda and Chandagiri.

CHAPTER 22

Nayaks of the Deep South

With the crumbling of the Vijayanagar empire, the Nayaks, representing Hindu rule in its dying phase, receded to their capital city of Madurai deep in the South to escape the wrath of the Muslim invaders. The more orthodox Hindu craftsmen followed them, seeking refuge in the fading glory of their feudal lords.

The Temple Becomes a Fort

With the growing uncertainties of temporal power, the building of new architectural masterpieces to the glory of God was hardly possible. There was little time, inspiration, or spirit for the blossoming of fresh artistic ideas. Building craftsmen had even under the aegis of the powerful Vijayanagar empire, kept themselves busy in erecting hall after hall of myriads of columns. They were quite content now to merely elaborate on the theme. The requirements of the keepers of the temple also specified buildings more of a defensive and essential character rather than of an innovative nature. It was more practical to protect the deity from the defiling invasions of Islam, than to attempt to create new exuberant forms of its house. The temple builders' art was virtually reduced to throwing a series of sacrosanct battlements around the divine presence, punctuated by well guarded gateways at the cardinal points.

Thus, the great temple of Srirangam near Trichurapalli (*Fig 22.01*) acquired several concentric rings of growth. Not merely over a decade or a century, but more like a town and a city, over a period of more than 500 years. During the Cholas' time it was probably no more than a village shrine consisting of the usual cella and *mandapa*. For various reasons it gained great religious popularity. The space within the walled enclosure surrounding the cella had become too small for conducting group ceremonies and accommodating the growing number of pilgrims. Nevertheless, both for sacred and practical reasons, it was not to be pulled down. The necessary additional space could have been provided by a courtyard or hall outside the front entrance. This, however, would have placed the new ancillaries at an inconvenient distance from the inner shrine. A more reasonable solution was to throw another concentric wall around the existing one. New structures could now be built in the space between the two parallel walls. This ensured that the additional halls were conveniently located more or less equidistant from the old inner shrine.

The Fort Becomes a City

Over a period of time with the accretion of greater religious merit and pilgrims arriving in thousands to attend festivities stretching over days, the temple precincts needed to be enlarged further. The precedents of the methodology of growth had now been established. More space was acquired and yet another even larger concentric

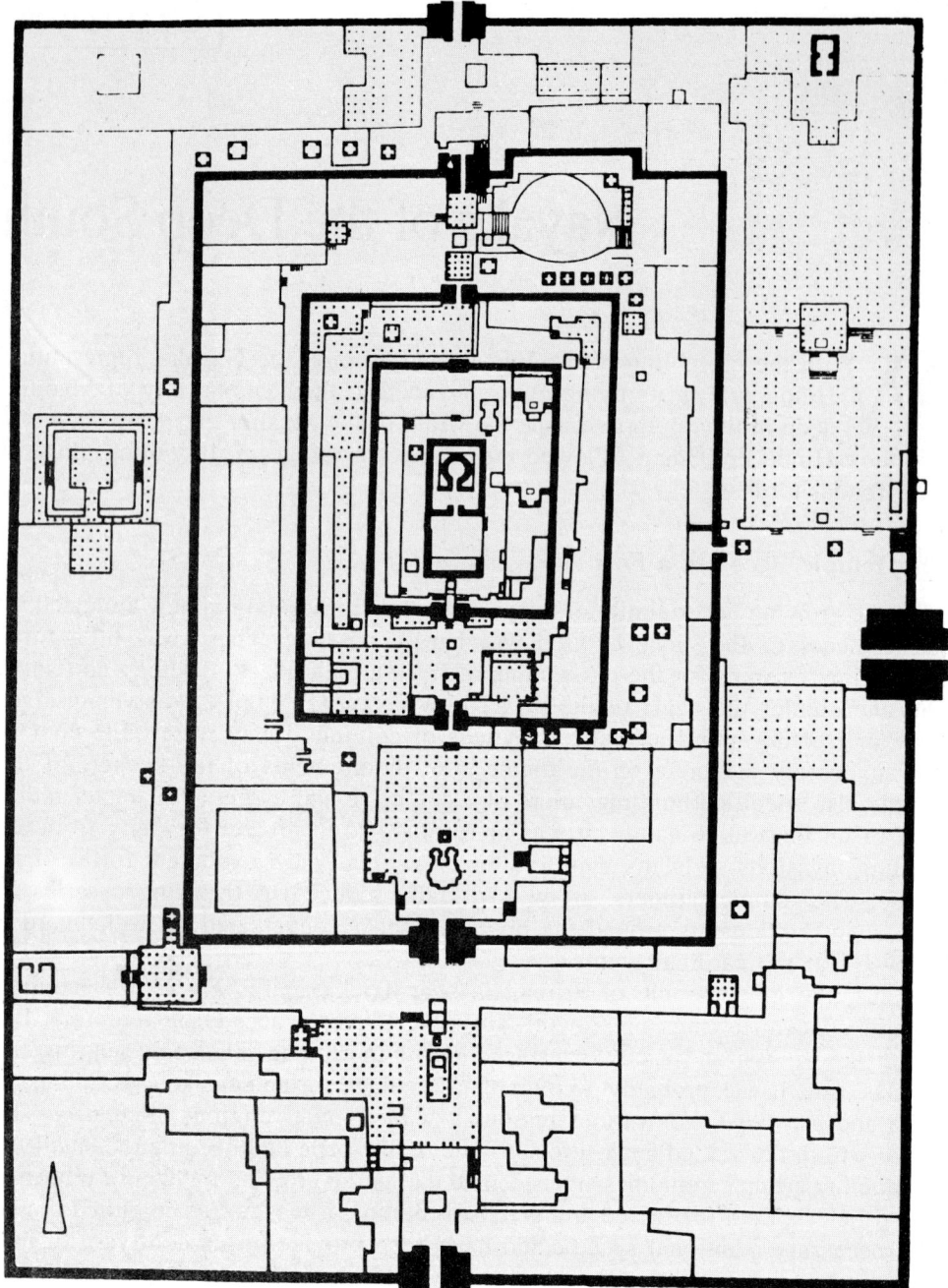

Fig 22.01 Plan of the temple at Srirangam that acquired several prakramas and concentric walls over a period of hundred of years to grow into a temple city

wall came up. With the many elaborate rituals to be performed in the vicinity of the central shrine, the original open court surrounding the *garbhagriha* was completely roofed over. The sacred image of the presiding deity that in any case could be only dimly perceived in the uncertain light of myriads of lamps, acquired an even more mysterious and sacred quality (*Fig 22.02*).

More people were attracted to the wonder temple. Sacred tanks were built, for them to bathe themselves in before confronting their gods. Two of these, one a conventional rectangular one representing the sun, and a semicircular one representing the moon, became part of another courtyard of the ever growing temple of Srirangam. The staff attached to such a temple naturally increased manifold, and a large part of it lived in the many houses built within the many temple walls.

More space was added by the well established pattern; another wall, another set of *gopurams*, more halls and consequently longer corridors, scores of rooms, cella and chambers were built as dictated by their functional use rather than architectural harmony. The only two elements that conformed to an architectural pattern in this great crescendo of haphazard growth were the surrounding wall and *gopurams*. The former were inevitably concentric to each other, and the latter invariably built along the cardinal axes. Over the years, the core of the Srirangam temple enclosure of 80 × 240 ft (24 × 73 m) had grown into a massive temple city of 2880 × 2475 ft (878 × 754 m) sprawling over an area of more than a quarter square mile. Ultimately, the rather fortuitous design of a temple as a series of concentric fortifications adorned with massive *gopurams* became an accepted style of temple building.

Fig 22.02 *The mysterious and sacred inner sanctum of the great temple city great temple city*

The Minakshi Temple at Madurai

Thus, the Minakshi temple, in Madurai (*Fig 22.03*), built in a few decades was, right from its inception, designed as a series of concentric courtyards or *prakramas* as they came to be classified. The ever alert Brahmin keepers also seized this opportunity to protect their precious god with rings of receding sanctity and security. Spaces around the shrine, the new *prakramas* were endowed with diminishing religious value directly proportional to their distance from the deity in the inner shrine. The outermost circle accommodated edifices of a practical rather than spiritual nature—accounts offices, dormitories for pilgrims, kitchens, shops dealing in essential items for rituals, maintenance workshops, and parking areas for the growing number of wooden festive chariots. The lower class menials who were needed to service these areas, were permitted only into the outlying areas.

Fig 22.03 Plan of the Meenakshi temple at Madurai enclosing two shrines within its prakrama (AD 1600)

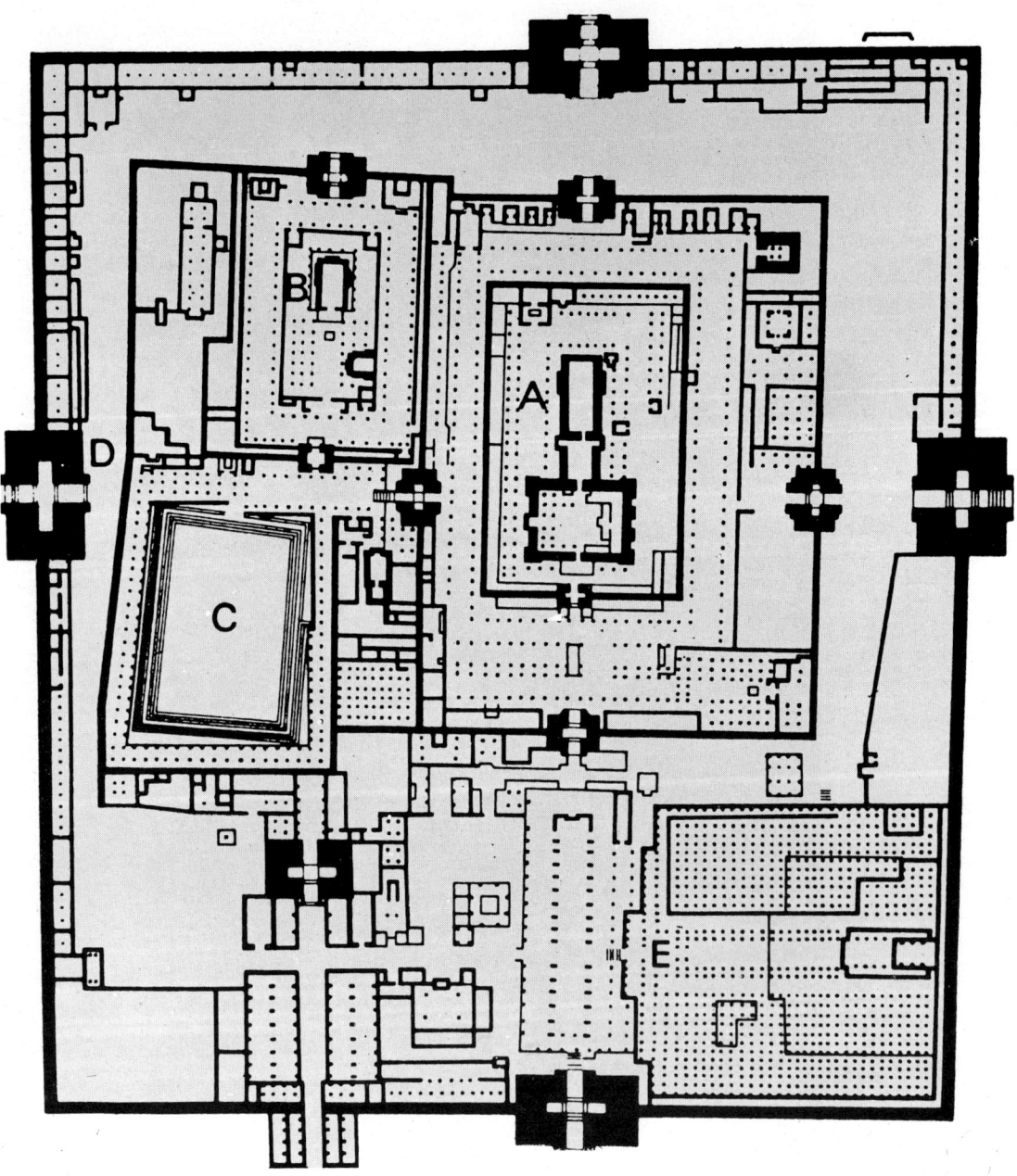

The inner *prakramas* contained pavilions for devotional singing and story telling, bathing tanks for ritual ablutions, and guest houses for important visitors. To such enclosures only caste Hindus were allowed. In the innermost courts were the kitchens of the Brahmins, pavilions for the dancing girls and the treasury, over both of which the Brahmin kept a watchful eye. Admittance to these was often restricted to the upper caste only. Even the chosen few allowed to have *darshan* of the deity, could wear only a brief loin cloth and nothing else—likely to make it impossible for them to hide away and pilfer the many valuable treasures of the temple within the folds of their clothes. The actual cella containing the graven image of the god was, of course, open only to the officiating priest and none else—not even royalty.

Halls of Many Pillars

With temple building beginning to approximate more to city and town planning, the building craftsman's art was restricted to the erection of pavilions, halls and *gopurams*, the subsequent ones larger and more gracious than the preceding.

The largeness of one of the flat-roofed halls, commissioned by Vinayak Mudali, a minister of the Nayak kings, necessitated the setting up of an elaborate stone cutters' factory. Column upon column and beam upon beam of a standard design was unendingly churned out of this factory. The hall situated within the Minakshi temple (*Fig 22.04*) complex needed 985 pillars to support its roof

Fig 22.04 A part of the hall of thousand pillars of the temple of Madurai

measuring 240 × 250 ft (73 × 76 m). This is the famous 'hall of thousand pillars' and little else. Today, ironically enough, the hall, surely one of the more arid products of Indian craftsmanship is a museum of drawings and photographs of the entire gamut of 1200 years of temple architecture of the South. Evidently, the builders and the patrons were undaunted by the monotony of their productions, drawing consolation from the massive size of their building ventures. They took seven years to build another, even larger, but less distinguished 330 ft (110.5 m) deep Choultri; this time at the command of the last Nayak, Tirumalai, who had it built outside the main precinct of the Madurai temple, south of the outermost *gopurams*.

The Soaring Gopurams

Even the *gopuram* had in many ways become a rather mechanical and repetitive form, growing only in size. The potential of its large surfaces, however, was exploited to some extent by the indefatiguable Indian sculptor. In its earlier stages, the profile of the *gopuram* was a rigid pyramidal tower, its horizontal tiers clearly defined by a series of vertical pilasters and horizontal cornices. Inevitably, however, the Hindu sculptor infused into it a more plastic quality, by making its profiles curvilinear (*Fig 22.05*). Subsequently, the pilasters and cornices lost their sharpness and became absorbed into one large amorphous jumbled mass of sculpture. The outermost, more than 150 ft (48 m) high *gopurams* of the Minkashi temple are the finest examples of this type of form. Once completed in the seventh century, the temple city of Madurai became the model for temple buildings in the South. Many old village shrines expanded into temple cities. As a result, there is hardly a temple of any religious consequence in the South which has not been surrounded by *prakramas*, entered through a series of axially aligned *gopurams*. Though evolving on much the same pattern as Srirangam, some acquired special distinctive features of their own.

The Corridors of Rameshwaram

Rameshwaram, situated at the tip of a narrow strip of land jutting out into the sea, is a maze of generously proportioned pillared verandas. The temple is not only surrounded by corridors (*Fig 22.06*) but is connected to the outer entrances by transverse covered passageways. To build these, masons patiently crafted and erected massive square columns sufficient to support roofing for these avenues, aggregating more than 3000 ft (914 m) in length. Rameshwaram has thereby acquired the dubious architectural record of possessing the longest corridors in the world (*Fig.22.07*). The Jambukeswar, within a mile of Srirangam, on the other hand, has probably the most massive and elaborate shafts and brackets for its numerous columns. These are more like elaborate inverted pyramids than mere pillars. The Shiva temple at Chidambaram boasts of a central shrine that is in the shape of a chariot, while the temple at Srivilliputtur is entered through a *gopuram* of 14 distinct storeys soaring over 200 ft (61 m) into the skies.

In spite of the many individual variations and gimmicks and world records, the essence of Hindu temple architecture in the South from the sixteenth to nineteenth century is contained in a series of concentric walls, *gopurams* of varying size, and halls and corridors of thousands of columns. Even the impressive skyline of *gopurams* of ascending heights is due more to fortuitous circumstances than planned strategy, unlike the preconceived intention of the pylons of Egyptian temples which they resemble.

Fig 22.05 The great curvilinear gopuram of Madurai soaring almost 50 m into the sky (AD 1600)

As we have seen, Indian architecture, for over a period of 300 years under the last Hindu lords of the South, did not show even a glimpse of a new form, a new idea. The keepers of the temples were content to surround *prakrama* with *prakrama* filled haphazardly with pillared halls connected by tedious corridors, relieved here and there by *gopurams*. Mercifully, this monotonous building activity could not be carried out endlessly. Whether the spread of Islam or the bankruptcy of Brahmin thought brought about the end, is a moot point; most likely it was a bit of both since the southernmost tip of India never really passed under direct Muslim control, nor has the Brahmin, even to date, quite lost his grip over the Hindu mind. Architecturally, however, the Hindu temple of the eighteenth and nineteenth century had become literally a fortress, that had drawn up the bridge to new ideas.

Building craftsmen, though, had not lost their intrinsic skills. Rather, the life-giving sap of new challenges had dried up, since the priest was content with altogether stereotyped forms. With Hindu temporal power also at a low ebb, patronage for the building of glorious monuments was no longer at hand. A number of orthodox Hindu craftsmen must have remained loyal to their religion. Some were occupied in renovating and refurbishing old Hindu temples, and some in building small village shrines here and there in the countryside. A good many others, either more loyal to their craft than to religious dogma, or through

◄ *Facing page*
*Fig 22.06 A portion of the
more than 1000 m length of
corridors, that are the major
feature of the seventeenth
century temple at
Rameshwaram*

compulsion, took the new challenge: that of building monuments for the Muslims. The North, which centuries earlier had succumbed to Muslim domination, was already bustling with activity. Given another set of rulers, another concept, the Indian craftsman set gamely about his task of erecting palaces, tombs and mosques.

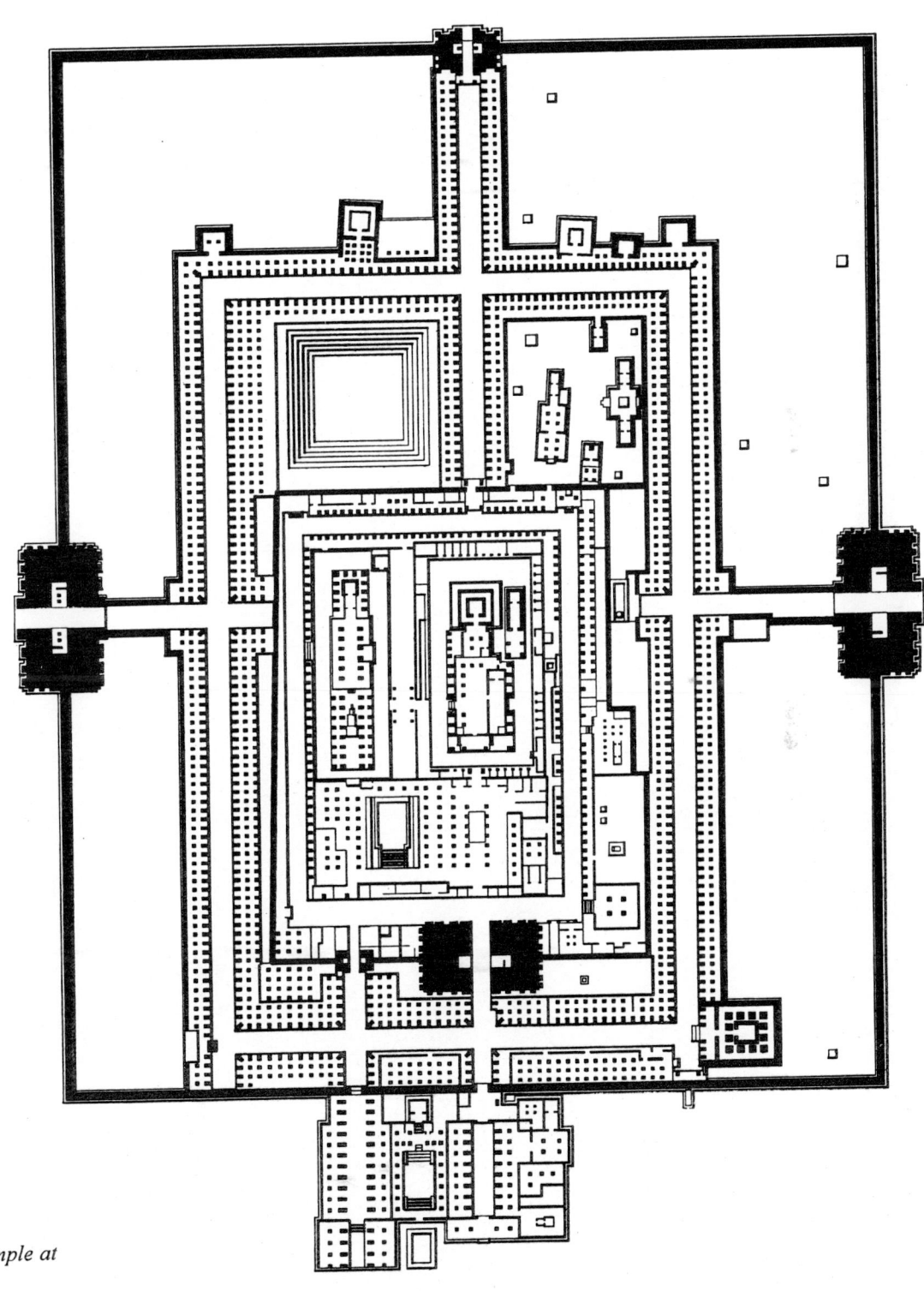

*Fig 22.07 Plan of the temple at
Rameshwaram*

Over the ensuing centuries, he convincingly proved that his craft was an undying one. It continued to flourish just as much under Muslim as it had under Hindu hegemony. The Muslim period, though, is another era in the history of the building arts of India.

Bibliography

BASHAM, A.L., *The Wonder That Was India*, Sidgwick & Jackson, London, 1954.

BATLEY, C., *Design Development of Indian Architecture*, John Murray, London, 1934.

BROWN, P., *Indian Architecture (Buddhists and Hindu Period)*, Taraporevala & Sons, Bombay, 1965.

CUNNINGHAM, A., *Archaeological Survey of India*, Vols. I–XXIII, Simla, Calcutta, 1903-30.

EDWARDES, M., *Indian Temples and Palaces*, Paul Hamlyn, London, 1959.

FERGUSSON, J., *History of Indian and Eastern Architecture*, John Murray, London, 1910.

GOETZ, H., *Five Thousand Years of Indian Art*, Methuen, London, 1959.

HAVELL, E.B., *The Ancient and Medieval Architecture of India*, John Murray, London, 1915; and *The Ideals of Indian Art*, John Murray, London, 1920.

KRAMRISCH, S., *The Hindu Temple*, University of Calcutta Press, Calcutta, 1946.

LAL, K., *Temples and Sculptures of Bhubaneswar*, Arts & Letters, New Delhi, 1970.

MARSHALL, J., *Annual Reports of the Archaeological Survey of India*, Calcutta, 1903-30, and *Mohenjodaro and the Indus Civilization*, London, 1931.

NEHRU, J., *The Discovery of India*, Meridian Books, London, 1946.

PIGGOT, S., *Pre-historic India*, Penguin Books, Harmondsworth, 1950.

RAWLINSON, H.G., *India: A Short Cultural History*, The Cresset Press, London, 1937.

ROWLAND, B., *The Art and Architecture of India*, Penguin Books, Harmondsworth, 1953.

THAPAR, R., *A History of India*, Penguin Books, Harmondsworth, 1966.

VOLWAHSEN, A., *Living Architecture of India*, MacDonald, London, 1970.

WHEELER, R.E.M., *Early India and Pakistan*, Thames and Hudson, London, 1959; and *The Indus Civilization*, Cambridge University Press, Cambridge, 1953.

ZIMMER, H., *The Art of Indian Asia*, Oxford University Press, London, 1968.

Glossary

amalaka: a flat fluted disc-like stone representative of the holy fruit "amala", usually at the summit of the shikhara.

antevasika: a resident in a hostel.

anda: literally "egg", the hemispherical dome of the stupa.

antralay: entrance chamber in front of shrines.

ardha mandapa: chamber before the main "mandapa" or hall.

aryaka: a Buddhist worshipful column.

asana: a sitting platform, bench or throne.

bazaar: market, shopping street.

bhog mandapa: hall in a temple where food from donors is supposed to be blessed by the enshrined deity.

Brahma: the Hindu god Supreme.

chaitya: from "chita", a pyre, but later a Buddhist sanctuary and ultimately the Buddhist hall of worship.

chaultri: pillared dormitory halls outside Dravidian temple.

chattra: literally umbrella; a honorific canopy at the apex of the Buddhist stupa.

chuna: lime, from the Chunar quarries in Bihar, the chief north Indian source of building lime.

dagoba: Simhalese terminology for the Buddhist stupa.

deul: from deva—god, signifying the inner sanctum, but in Orissa applicable to the temple as a whole.

garbhagriha: literally "the womb", the most sacred inner sanctum of the temple where the deity is enshrined.

ghat: platforms or steps at edge of lake or river water.

ghats: popular term for the mountainous territory south of the Deccan Plateau and in the peninsular portion between the Bay of Bengal and the Arabian sea.

gopuram: monumental South Indian temple gateway.

harmika: the square platform at the apex of the stupa sometimes surrounded by a railing and from top of which the honorific umbrella rises.

hinayana: the earliest form of Buddhism that did not include the worship of images of the Buddha as a god.

Indra: the ideal warrior god of the Hindu pantheon.

jagmohan: Orissan terminology for hall in front of the sanctuary.

Jina: mythological reformer in the Jain religion.

Kailasa: the haven of the Hindu god Shiva.

kalasa: literally vase, religious inverted not installed over the amalaka as a finial for the shikhara.

kalyan mandapa: hall of "marriage of the gods" in South Indian temples.

lat: pillar or column.

mahal: palace, hall.

mahayana: later development of Buddhism that accepted Buddha as saviour and permitted the worship of its image.

mandapa: large hall, generally the chambers preceding the inner sanctum of the Hindu temple.

mandir: temple.

medhi: terrace or platform.

nandi: the sacred bull.

nat mandir: dancing hall in Hindu temple, particularly in Orissan terminology.

navranga: central hall of temple, particularly in Gujarat temples.

panchayatna: temple consisting of a central shrine and four subsidiary shrines at the corners.

pradakshinapath: processional ambulatory passage, particularly Buddhist, around the stupa.

prakrama: open courtyard, particularly the ring of courtyards in South Indian temples.

purana kalasa: symbolic vase of plenty, pot and foliage motif of column capitals of the Gupta period.

ratha: literally car or chariot; particularly car used in Hindu processional festivals, sometimes also signifying temple.

sabha mandapa: assembly hall.

santhagara: village or town meeting hall.

seni: guilds.

shikhara: literally mountain peak but generally the spire or tower over north Indian Hindu temple.

silpa sastra: ancient Indian treatise on architecture, building and allied arts.

stambha: column or pillar.

sthapati: master craftsman.

stupa: originally a folk funerary mound but crystallised by Buddhism into a hemispherical form to enshrine a sacred spot or relic.

Surya: the Hindu Sun god.

suryavanshi: Rajput tribes claiming descent from the Sun god.

tirathankara: Jain religious reformers.

torana: gateway, particularly the ceremonial gates at the cardinal points of the stupa.

trimurti: sculpture embodying three gods, particularly Brahma, the creator; Vishnu, the preserver; and Shiva, the destroyer.

urusringa: a half profile shikhara superimposed over main shikhara, particularly in the Khajuraho style.

vastu sastra: ancient Indian treatise on the "rules of architecture."

vedas: ancient texts embodying the essence of the Aryan religious life in the form of mantras or chants to be recited at religious ceremonies.

vedi: altar.

vedica: Buddhist railing, generally demarcating a sacred area.

vihara: residence for Buddhist or Jain monks.

vimana: pyramidical tower over the inner sanctum, generally of South Indian temples.

Index